Mastering Snowflake DataOps with DataOps.live

An End-to-End Guide to Modern Data Management

Ronald L. Steelman, Jr.

Foreword by Doug Needham,
Principal Solutions Architect, DataOps.live

Apress®

Mastering Snowflake DataOps with DataOps.live: An End-to-End Guide to Modern Data Management

Ronald L. Steelman, Jr.
Justin, TX, USA

ISBN-13 (pbk): 979-8-8688-1753-3 ISBN-13 (electronic): 979-8-8688-1754-0
https://doi.org/10.1007/979-8-8688-1754-0

Managing Director, Apress Media LLC: Welmoed Spahr
Acquisitions Editor: Shaul Elson
Development Editor: Laura Berendson
Coordinating Editor: Gryffin Winkler

Distributed to the book trade worldwide by Springer Science+Business Media New York, 1 New York Plaza, New York, NY 10004. Phone 1-800-SPRINGER, fax (201) 348-4505, e-mail orders-ny@springer-sbm.com, or visit www.springeronline.com. Apress Media, LLC is a Delaware LLC and the sole member (owner) is Springer Science + Business Media Finance Inc (SSBM Finance Inc). SSBM Finance Inc is a **Delaware** corporation.

For information on translations, please e-mail booktranslations@springernature.com; for reprint, paperback, or audio rights, please e-mail bookpermissions@springernature.com.

Apress titles may be purchased in bulk for academic, corporate, or promotional use. eBook versions and licenses are also available for most titles. For more information, reference our Print and eBook Bulk Sales web page at http://www.apress.com/bulk-sales.

Any source code or other supplementary material referenced by the author in this book is available to readers on GitHub. For more detailed information, please visit https://www.apress.com/gp/services/source-code.

If disposing of this product, please recycle the paper

This book is dedicated to those who think or who have been told that they can't. Remember this: You can!

Table of Contents

About the Author xv

About the Technical Reviewer xvii

Foreword xix

Chapter 1: DataOps 1

How Are DataOps and DevOps Related 2

Evolution of DataOps 3

The Rise of Big Data and the Need for Agile Data Development 4

From DevOps to DataOps: Bridging the Gap 7

Automation and CI/CD for Data Pipelines 9

The Emergence of DataOps Platforms 12

Data Quality, Governance, and Security in the DataOps Era 14

The Future of DataOps: AI and ML Integration 17

Principles of DataOps 19

Continuous Integration and Continuous Delivery (CI/CD) for Data 20

Collaboration and Communication in Data Teams 21

Automation and Orchestration in Data Management 23

Data Quality and Governance 24

Scalability and Flexibility in Data Operations 26

Benefits of Adopting DataOps 27

The DataOps Process 28

DataOps Pipelines for Data 29

Conclusion 31

Chapter 2: Pillars of True DataOps.. 33

Pillar 1: ELT (and the Spirit of ELT) 34

Pillar 2: Continuous Integration/Continuous Delivery (CI/CD)........................ 35

Pillar 3: Component Design and Maintainability 36

Pillar 4: Environment Management.. 38

Pillar 5: Governance and Change Control.. 39

Pillar 6: Automated Data Testing and Monitoring 40

Pillar 7: Collaboration and Self-Service 41

Conclusion 42

Chapter 3: MLOps 43

What Is MLOps? 43

The MLOps Lifecycle.. 45

1. Problem Definition and Business Understanding 46

2. Data Collection and Preparation.. 46

3. Feature Engineering and Selection 47

4. Model Development and Training.. 47

5. Model Evaluation and Validation... 47

6. Model Deployment... 47

7. Monitoring and Maintenance.. 48

8. Retraining and Model Iteration 48

9. Governance and Compliance.. 48

10. Collaboration and Feedback... 48

Core Principles of MLOps.. 49

Challenges in Implementing MLOps 50

Versioning in MLOps: Models, Code, and Data 52

Automation in MLOps.. 54

Model Deployment Strategies... 56

Monitoring and Maintenance of ML Models.. 58

MLOps for Governance and Compliance 59

Case Studies and Real-World Applications 60

Conclusion 62

Chapter 4: DataOps Best Practices 65
Establishing Cross-Functional Teams for Data Collaboration 65
Collaboration Tools for Enhancing Cross-Functional Teamwork 67
Automating Data Pipelines for Consistency and Speed 68
Data Quality and Testing: Ensuring Reliability at Scale 69
Managing Data Governance and Compliance in DataOps 71
Version Control and Data Lineage: Tracking Changes Effectively 72
Scaling DataOps with Cloud and Hybrid Architectures 74
Conclusion 75

Chapter 5: Understanding Snowflake 77
Overview of Snowflake Architecture and Features 77
Key Concepts of Snowflake 79
Benefits of Snowflake for Data Warehousing 80
Snowflake's Multi-Cluster Shared Data Architecture 82
Snowflake's Unique Storage and Compute Separation 83
Snowflake's Data Sharing Capabilities 84
Performance Optimization in Snowflake 85
Snowflake for Data Security and Compliance 86
Integrating Snowflake with Third-Party Tools and Platforms 87
Use Cases and Industry Applications of Snowflake 88
Conclusion 91

Chapter 6: Introduction to Git 93
What Is Version Control? 93
Why Is Version Control Important? 94
What Is Git? 95
Understanding Git's Core Principles and Workflow 96
Best Practices for Using Git in Collaborative Development 98
Conclusion 99

Chapter 7: Getting Started with Git....... 101

Basic Git Concepts....... 101

Installing Git....... 103

Working with Git....... 107

Collaboration with Git....... 108

Aliases....... 109

Tagging....... 110

Remotes....... 111

Conclusion....... 112

Chapter 8: Advanced Git Topics....... 113

Branching....... 113

Merging....... 115

Best Practices....... 116

Troubleshooting and Tips....... 118

Conclusion....... 119

Chapter 9: Introduction to DBT....... 121

What Is DBT?....... 121

Key Concepts in DBT....... 123

DBT Workflow and How It Fits into Data Pipelines....... 125

Setting Up DBT....... 127

Developing and Running DBT Models....... 129

Testing and Documentation in DBT....... 131

Testing in DBT....... 131

Documentation in DBT....... 133

Testing and Documentation Together....... 134

Version Control and Collaboration with DBT....... 134

Conclusion....... 136

Chapter 10: Advanced DBT Techniques and Best Practices 137

Optimizing DBT Models for Performance 138

Advanced Data Transformations with DBT 140

Customizing DBT with Macros and Jinja 143

Leveraging DBT's Advanced Testing Capabilities 146

Managing Large-Scale DBT Projects 149

Optimizing DBT Run-Time: Best Practices 151

Using DBT for Incremental Loading 153

Versioning and Collaboration in Advanced DBT Projects 155

DBT Artifacts: Understanding Logs and Artifacts for Debugging 157

Scaling DBT for Enterprise-Level Deployments 160

Conclusion 161

Chapter 11: Introduction to the DataOps.live Platform 163

About the DataOps.live Platform 163

Key Features and Capabilities 164

Development Principles 165

Use Cases: DataOps.live in Snowflake Data Cloud 167

Conclusion 168

Chapter 12: DataOps.live and DataOps: Better Together 169

Integration of DataOps Principles into DataOps.live 169

How DataOps.live Enhances DataOps Practices 171

Case Studies: Illustrating the Synergy Between DataOps and DataOps.live 172

Conclusion 175

Chapter 13: Essential Elements of DataOps.live 177

Accounts 177

Groups and Subgroups 178

Projects 179

Pipelines and Jobs 181

Conclusion 182

Chapter 14: Getting Started with DataOps.live 183

Acquiring DataOps.live from Snowflake Marketplace 183

Setting Up Your DataOps.live Environment 184

Integrating DataOps.live into Your Data Workflow 185

Managing Data Pipelines and Jobs in DataOps.live 185

Scaling and Monitoring Your DataOps.live Projects 186

Best Practices for Getting the Most Out of DataOps.live 186

Conclusion 187

Chapter 15: Managing Your Environments 189

Understanding the Key Types of Environments 189

Best Practices for Environment Segregation and Configuration 190

Creating DataOps Environments 191

Branching Strategies 193

Conclusion 196

Chapter 16: Build Your First DataOps Pipeline 199

Running Your First Pipeline 199

Conclusion 202

Chapter 17: Getting Started with SOLE 203

Introduction to SOLE and Data Products 203

Setting Up SOLE for Data Products 205

Modular and Self-Describing Object Configuration 207

Best Practices for Organizing SOLE Configurations 209

Advanced Configuration Options and Use Cases 211

Conclusion 213

Chapter 18: Getting Started with MATE 215

MATE Versus DBT 215

Core Concepts of MATE 216

Building Models with MATE 217

Using DBT Packages with MATE 219

Macros with MATE 221
MATE Best Practices 222
Conclusion 224
Chapter 19: Managing Multiple Databases with DataOps.live 225
Understanding Database Management in DataOps.live 226
Setting Up Multiple Databases in DataOps.live 227
Optimizing Cross-Database Workflows 228
Best Practices and Troubleshooting 229
Conclusion 231
Chapter 20: DataOps.live Orchestrators 233
Data.World Orchestrator 234
VaultSpeed Orchestration in DataOps.live 236
Conclusion 239
Chapter 21: Build Only Changed Models 241
Building Change-Only Models 241
Chapter 22: DataOps.live REST API 243
Chapter 23: Medallion Architecture 247
Understanding Medallion Architecture: Key Concepts and Principles 247
The Bronze Layer: Raw Data Ingestion and Storage 250
Transforming Data in the Silver Layer: Cleaning and Enrichment 252
The Gold Layer: Data Aggregation and Analysis for Business Insights 254
Implementing Medallion Architecture with DataOps.live 256
Medallion Architecture Best Practices and Optimization Strategies 257
Scaling Medallion Architecture for Large Data Environments 260
Use Case: Medallion Architecture in Solar Energy 261
Conclusion 264

Chapter 24: Kimball Architecture 265
Overview of Kimball Architecture: Key Principles and Concepts 265
Designing Dimensional Models: Facts and Dimensions 267
Building Data Marts: Organizing Data for Business Use 269
ETL Processes in Kimball Architecture 270
Kimball Architecture Best Practices and Optimization Strategies 272
Conclusion 273

Chapter 25: DataVault 2.0 Architecture 275
Introduction to DataVault 2.0 Architecture 275
Key Principles of DataVault 2.0 277
Core Components: Hubs, Links, and Satellites 279
DataVault Hubs 280
DataVault Links 282
DataVault Satellites 285
Benefits of DataVault 2.0 for Modern Data Environments 287
Implementing DataVault 2.0 with DataOps.live 289
Best Practices for DataVault 2.0 Modeling 290
Use Case: DataVault 2.0 in the Healthcare Industry 292
Conclusion 295

Chapter 26: Combining Medallion with DataVault 2.0 and Kimball 297
Combining Medallion, DataVault 2.0, and Kimball Architectures 298
DataVault 2.0 in the Silver Layer 299
Using Kimball in the Gold Layer 300
Best Practices for Integrating Datavault 2.0 with the Medallion Architecture 302
Optimizing Data Flow from Bronze to Gold: A Unified Approach 303
Managing and Scaling the Combined Architecture 306
Use Case: Combining Medallion, DataVault 2.0, and Kimball in the Finance Industry 307
Conclusion 310

Chapter 27: Entity-Relationship (ER) Modeling and Beyond ... 311

Fundamentals of ER Modeling: Entities, Attributes, and Relationships ... 311

Advanced ER Modeling, Cardinality, Normalization, and Integrity Constraints ... 313

Extending ER Models: Incorporating Hierarchies, Temporal Data, and Beyond ... 314

From ER Models to Modern Data Architectures: Bridging the Gap ... 316

Conclusion ... 317

Chapter 28: Event-Driven Data Models ... 319

Understanding Event-Driven Architecture: Principles and Core Concepts ... 319

Designing Event-Driven Data Models: Events, Streams, and State Changes ... 321

Integrating Event-Driven Models with Data Warehouses and Data Lakes ... 323

Best Practices for Implementing Event-Driven Data in Modern Architectures ... 324

Conclusion ... 326

Chapter 29: Graph Data Modeling ... 327

What Are Graph Data Models: Structure and Key Concepts ... 327

Designing Graph Databases: Nodes, Edges, and Properties ... 329

Advanced Graph Querying: Traversals, Patterns, and Algorithms ... 331

Conclusion ... 333

Index ... 335

About the Author

Ronald L. Steelman, Jr. is a seasoned expert with over 20 years of experience in data, analytics, artificial intelligence (AI), and software development. Throughout his career, Ronald has worked across various industries, from large enterprises in healthcare and technology to nimble startups and consulting firms. His broad professional background has enabled him to develop a unique perspective on the challenges and opportunities organizations face regarding data management and analytics. With his deep understanding of cutting-edge technologies and methodologies, Ronald has consistently delivered innovative solutions that help businesses stay ahead in a rapidly evolving digital landscape.

Specializing in data integration, AI/ML (machine learning), analytics, and DataOps, Ronald has become an expert in the field and a thought leader in the Snowflake ecosystem. His expertise is creating streamlined data processes that maximize efficiency while driving actionable insights. Over the years, he has helped countless organizations transform how they handle data, providing them with the tools and strategies they need to build robust, scalable infrastructures. Ronald's comprehensive understanding of DataOps methodologies allows him to simplify complex workflows, enabling teams to focus on what matters most: deriving value from their data.

This technical book was born from his desire to help readers integrate a DataOps strategy into their organizations and better utilize the DataOps.live platform to achieve their business goals. Ronald's goal with the book is to make sophisticated data processes more accessible, empowering professionals to take complete control of their data pipelines and deliver more efficient, automated solutions.

In addition to his practical experience, Ronald attended the University of Oklahoma, where he built a solid foundation in data and technology. His passion for learning and continuous growth is evident in his decision to author *Mastering the Snowflake SQL API with Laravel 10.*

Beyond his professional achievements, Ronald is a curious and adventurous individual with many personal interests. When he's not exploring the latest technological advancements, he enjoys traveling, cooking, and keeping reef and saltwater aquariums. He's also an avid reader, constantly seeking new knowledge in both his personal and professional life. His diverse interests reflect his belief in a balanced approach to life, combining technical expertise with creativity and exploration.

About the Technical Reviewer

Dayakar Siramgari brings over two decades of innovation in cloud architecture, data engineering, and GenAI platforms. Currently at T-Mobile, he leads transformative efforts in digital modernization, cost optimization, and enterprise-scale analytics. His cross-platform expertise spanning Azure, AWS, Snowflake, and Databricks enables him to architect future-ready systems across industries. With a unique blend of technical depth and visionary thinking, Dayakar's review ensures this book is both practical and cutting-edge.

Foreword

I met Ronald Steelman some time ago, shortly after I started working with DataOps.live. Since then, I have had the pleasure of witnessing his passion for elegant solutions applied at several customer engagements.

I have always believed that the most powerful tools are those that not only solve a technical problem but also fundamentally change the way we think and work. The DataOps.live platform is one such tool, and this book is the key to unlocking its full, transformative potential.

It is a genuine pleasure to introduce this work by Ronald Steelman. He has poured that same energy and passion for creating innovative and robust solutions into these pages, creating what I believe will become the foundational text for anyone working with our platform.

What makes this book so remarkable is its dual nature. It is both a panoramic overview and a microscopic examination. He masterfully explains the strategic importance of DataOps principles while simultaneously diving deep into the technical nuts and bolts of the DataOps.live platform.

You are holding more than a book; you are holding a repository of expert knowledge, a practical guide to execution, and a clear vision for the future of data management. I commend him for this outstanding achievement and wholeheartedly recommend it.

Doug Needham
Principal Solutions Architect
DataOps.live

CHAPTER 1

DataOps

At this point, you may be wondering, "What exactly is DataOps?" If you're asking yourself what exactly is DataOps, this book will take you through answering that question and use it as a foundation to build upon for implementing true DataOps processes. We will start by diving into the world of DataOps philosophy and the #TrueDataOps movement, which will equip readers with fundamental knowledge they will need to fully embrace the concepts and methodologies of DataOps to enable building robust and comprehensive DataOps workflows. While this book will focus on Snowflake and the DataOps.Live platforms, the concepts you will learn throughout can be applied as a general set of principles to any database solution. In fact, I encourage you to take what you learn from this book and explore new and innovative ways to apply it to other database solutions. DataOps has much growing to do, and like any toddler, it takes patience, persistence, lessons learned, collaboration, and thinking outside the box.

At its most basic understanding, DataOps, short for Data Operations, merges data engineering and operations to provide an automation-first approach to building and scaling data products. You can achieve this through practices, processes, and technologies. These culminate into best practices, methodologies, coding standards, and Continuous Integration/Continuous Development (CI/CD) pipelines. As a practitioner of DataOps, you'll be expected to encompass all of these concepts into a final solution that will ultimately help you automate building, testing, and deploying consistently and through embracing a DRY (don't repeat yourself) approach. If employed correctly, a proper DataOps approach can lead to a lean database engineering lifecycle and help bring to focus a more collaborative and streamlined approach to development operations.

Traditional database engineering has followed a fractured process that tends to lend itself to a more error-prone approach to building, maintaining, and evolving a robust data strategy. Even as you compare database engineering to the more typical software engineering, there are deviations in their approaches. It is important to remember that

R. L. Steelman, Jr., *Mastering Snowflake DataOps with DataOps.live*, https://doi.org/10.1007/979-8-8688-1754-0_1

when you talk about data, you are talking about a nebulous "thing" that is always in flux and continually growing. In contrast, with software engineering, things are very linear and controlled. Even more so, software engineering, most of the time, relies heavily on a solid data strategy. If anything, the latter is more straightforward than the former because software engineering principles are based on a reasonably linear concept: build, deploy, and watch as it thrives. It is this break from such a fractured process that DataOps has adopted a variant of the DevOps approach to the engineering lifecycle in the "infinite loop" of DataOps, as seen in Figure 1-1.

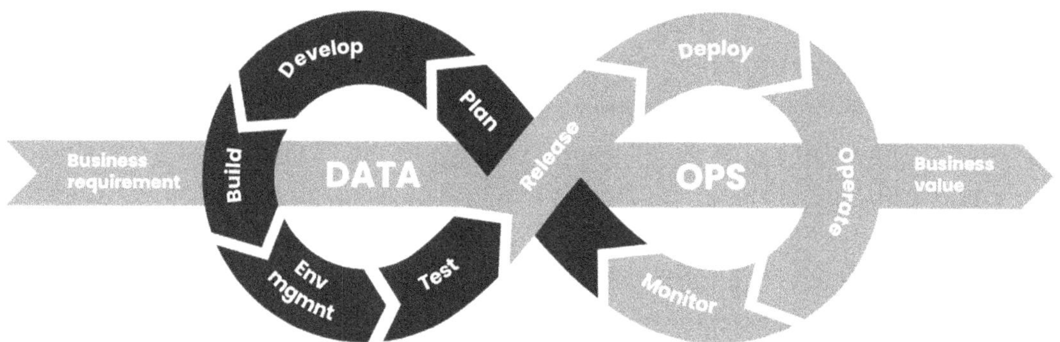

Figure 1-1. *The DataOps Infinite Loop (Credit: TrueDataOps.com)*

Up to this point, I hope you have a better conceptual understanding of DataOps, but what does all of this mean? Suppose you are that one individual who put this book down after reading the previous couple of paragraphs to do some trusty Google searches. In that case, you've probably already seen some terminology: GIT, version control, CI/CD, pipelines, etc. If this is you, that's great! You're already hungry for more. If that is not you, don't worry because that is what this book aims to do. My goal is to not only arm you with the knowledge you need to become a DataOps guru but also to give you the tools and technologies to make your life easier.

How Are DataOps and DevOps Related

Before I discuss DataOps in-depth, I want to discuss DevOps briefly. If you've been in the engineering field for any time, you've been exposed to the DevOps concept at least once. In fact, as you can see in Figure 1-2, DevOps has its own "infinite loop" representation that, at a high level, defines the approach.

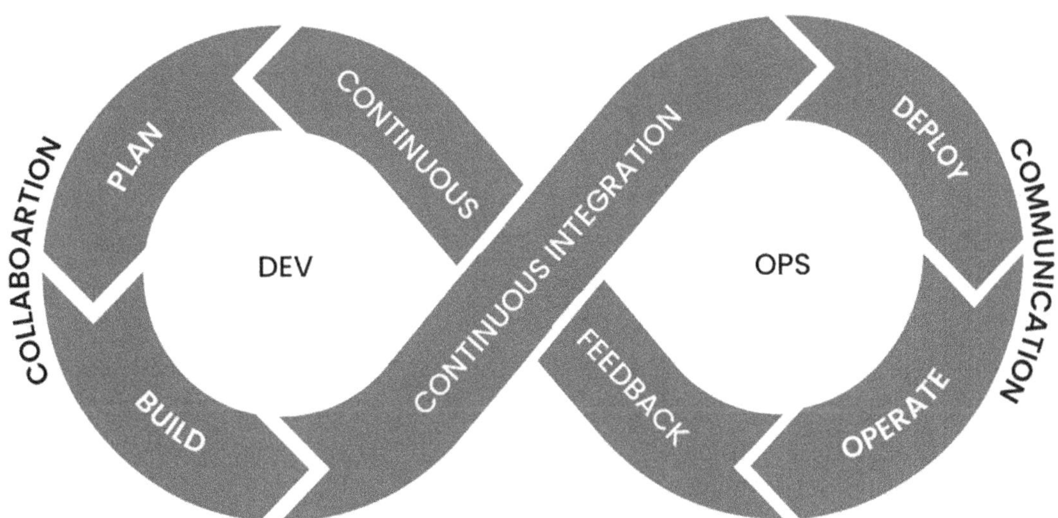

Figure 1-2. *The DevOps Infinite Loop*

As you can see from the DevOps Infinite Loop, the methodologies overlap. Because database development is complex and ever-evolving, the infinite loop is somewhat more granular to help manage those complexities. With this knowledge, you can boil this down into a few conceptually crucial differences. DevOps and I are very abstract, and we deal primarily with software development, feature upgrades, and break-fix deployments. On the other hand, DataOps goes a few layers deeper and focuses on the goals of data governance, resource management, data quality, data orchestration, data modeling, and data monitoring, and provides some bleed-over into data analytics. It is important to note that while there is some overlap in the world of analytics, DataOps still needs to encompass AI and machine learning (ML) fully. You can represent these through a similar process known as MLOps. We will touch on MLOps briefly later in this book.

Evolution of DataOps

As organizations have become more data-driven, the need for more efficient, scalable, and reliable data management practices has intensified. Traditional data handling methods—often siloed and reliant on manual processes—have struggled to keep pace with the growing demands for real-time insights, data quality, and governance. DataOps emerged as a solution to these challenges, offering a set of practices, technologies, and principles designed to streamline and automate data workflows, much like DevOps did for software development.

DataOps is more than just a methodology; it represents a cultural shift in how data teams collaborate with business units, developers, and IT. Its evolution is tied deeply to the rise of big data and the complexity of modern data ecosystems, where data pipelines need to be fast, flexible, and reliable. At its core, DataOps focuses on improving communication, integration, and automation across the entire data lifecycle—from ingestion and transformation to analytics and reporting. Doing so enables organizations to deliver value from their data more efficiently and with greater agility.

Over time, DataOps has evolved from a concept inspired by DevOps to a fully fledged operational framework supported by specialized platforms and tools. These developments have allowed businesses to scale their data operations while ensuring governance, security, and quality. As the need for real-time analytics, machine learning, and data-driven decision-making grows, DataOps will continue to play a crucial role in shaping the future of how organizations manage their most valuable asset—data.

The Rise of Big Data and the Need for Agile Data Development

The explosion of data in the early 21st century marked a pivotal moment for businesses and organizations across industries. With the rapid rise of digital platforms, social media, mobile devices, and IoT (Internet of Things), the volume, variety, and velocity of data generated began to grow at an unprecedented rate (Figure 1-3). Traditional data management systems, built around structured, transactional data, quickly became overwhelmed by the sheer scale of this "big data." Organizations struggled to manage the influx of information and the complexities that came with it.

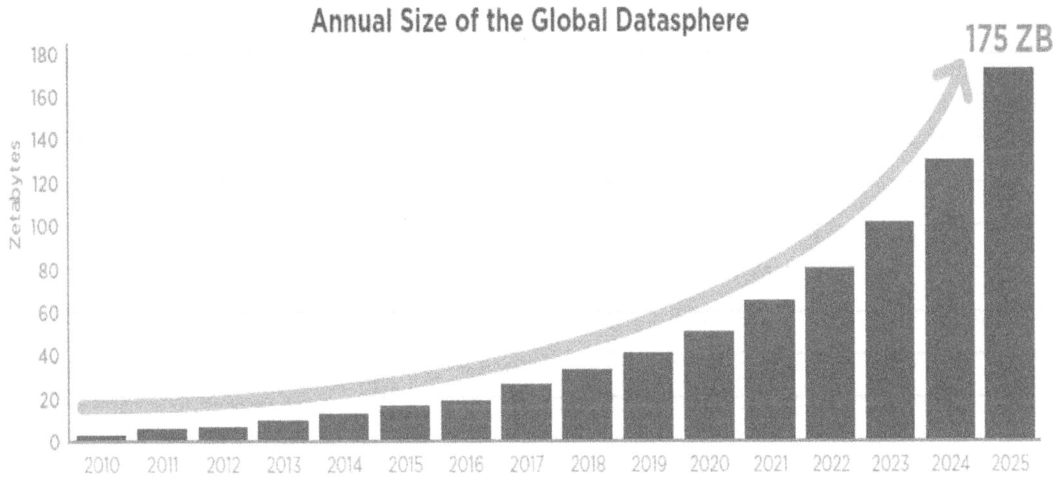

Figure 1-3. *Annual Size of the Global Datasphere (Credit: Forbes)*

In the early days, companies relied heavily on centralized data warehouses and static ETL (Extract, Transform, Load) processes to store and manage their data. These systems worked well for small- to medium-sized datasets, typically from limited sources like internal databases or enterprise applications. However, as more data sources began to emerge—ranging from social media interactions and sensor data to web traffic and customer reviews—the rigidity of these systems became a bottleneck. Data became siloed, and moving from ingestion to analysis took longer, hampering organizations' ability to make timely, data-driven decisions.

At the same time, the expectations for data usage were shifting. No longer was it enough to generate monthly or even weekly reports; businesses now needed real-time insights to respond quickly to market changes, customer behavior, and competitive pressures. This demand for faster, more flexible data processes exposed the weaknesses in the existing data infrastructure. The traditional waterfall approach to data development—where requirements were set in stone upfront, and teams would develop each pipeline stage in isolation—could no longer keep pace with the need for rapid iteration and adaptation.

Agile data development emerged as a solution to these challenges, borrowing principles from the Agile software development movement that had already transformed how IT teams worked. In Agile software development, cross-functional teams collaborate closely, iterating quickly through minor, incremental releases rather than working in isolation on massive, months-long projects. Applying these principles to data development meant rethinking how you build data pipelines, shifting from rigid, linear processes to more collaborative, iterative approaches. This allowed data teams to be more responsive to changing business needs and to incorporate new data sources or modify analyses on the fly.

One of the core drivers behind the move to Agile data development was the need for better alignment between data teams and business users. In traditional data management models, there was often a significant gap between the data engineers who built the pipelines and the analysts or decision-makers who used the data. This gap led to inefficiencies, with data often being delivered too late or in a format that was not useful. Agile practices helped bridge this divide by fostering more direct communication and collaboration, enabling quicker feedback loops and more relevant data outputs.

Another critical element of Agile data development is automation. In the early days of big data, many data processes were manual, requiring significant human intervention to clean, transform, and load data. These manual processes were time-consuming and prone to error, further delaying the delivery of insights. With the advent of new tools and platforms, automation became a central feature of Agile data development, allowing teams to streamline repetitive tasks and focus on higher-value work, such as data analysis and modeling.

Additionally, the need for agility in data development was driven by the evolving complexity of the data landscape. With data coming from an ever-increasing number of sources in various formats—structured, semi-structured, and unstructured—traditional methods for managing data could not keep up. Agile methodologies introduced a more flexible, modular approach to building data pipelines, enabling teams to quickly adapt to new data types and integration requirements without overhauling entire systems.

Cloud computing also played a critical role in supporting the rise of Agile data development. The scalability and flexibility offered by cloud platforms allowed organizations to move away from expensive, on-premise infrastructure and embrace more elastic, cost-effective solutions for managing big data. Cloud services made it easier to spin up new environments, test data processes, and scale resources up or down as needed, providing the agility to handle fluctuating workloads and real-time data needs.

Finally, the rise of big data underscored the need for a more dynamic approach to data governance and security. As data volumes grew and more stakeholders became involved in using data for decision-making, ensuring data quality and compliance with regulatory requirements became even more critical. Agile data development practices helped address these concerns by embedding governance and security measures throughout the data pipeline rather than treating them as afterthoughts. This enabled organizations to maintain control over their data assets while benefiting from Agile processes' speed and flexibility.

Because of these evolutions, the rise of big data fundamentally transformed how organizations manage and develop their data operations. Traditional, siloed approaches quickly proved inadequate in addressing the demands of a dynamic, data-driven environment. In response, Agile data development emerged as a more collaborative, flexible, and automated methodology, enabling teams to build and adapt data pipelines with incredible speed and efficiency. This shift has empowered organizations to manage their data's increasing scale and complexity and extract value from it with heightened precision and agility.

From DevOps to DataOps: Bridging the Gap

The evolution of DevOps in the early 2000s set the stage for a radical shift in how software development and IT operations collaborated to deliver applications and services more quickly and reliably. The DevOps philosophy emphasizes continuous integration, automation, and a strong alignment between development and operations teams. As businesses recognized the value of these principles, it became clear that you could apply the same approach to data management and analytics, which faced many of the same challenges—siloed teams, slow processes, and a lack of agility. This realization sparked the birth of DataOps, a methodology designed to bring DevOps principles to the world of data.

DevOps succeeded in breaking down the barriers between development and IT teams, leading to faster software releases and improved collaboration. However, data teams—comprised of data engineers, data scientists, analysts, and business users—were still mainly operating in isolation. DataOps arose to bridge the gap by applying the same collaborative, iterative, and automated principles that DevOps used in software development. In DataOps, the focus shifts from code delivery to data delivery, ensuring that data pipelines are built, tested, and deployed in an agile, automated manner.

One of the key lessons from DevOps that inspired DataOps is the idea of continuous integration and continuous delivery (CI/CD). In DevOps, CI/CD refers to the practice of automatically building, testing, and deploying code changes to ensure that software can be released quickly and reliably. DataOps takes this same concept and applies it to data pipelines. In traditional data environments, changes to data pipelines—whether adding new data sources, altering transformations, or deploying models—often involved manual processes that introduced delays and increased the risk of errors. By adopting CI/CD in data operations, DataOps enables data teams to automate the testing and deployment of data workflows, ensuring that changes are deployed faster and with greater accuracy.

Another important parallel between DevOps and DataOps is the focus on automation. DevOps relies on automation tools in software development to handle everything from code testing to infrastructure provisioning. Similarly, DataOps strongly emphasizes automating the data lifecycle, from data ingestion and transformation to quality checks and monitoring. Automation reduces the manual effort required to manage data pipelines and helps ensure data consistency and reliability across the organization. By automating repetitive tasks, DataOps teams can focus on more strategic, value-added activities, such as optimizing data models or generating insights from analytics.

Collaboration is another pillar of DevOps that has been successfully carried over to DataOps. DevOps promotes a culture of shared responsibility, where development and operations teams work together toward common goals, breaking down the traditional silos between them. DataOps brings this spirit of collaboration to data teams, encouraging cross-functional cooperation between data engineers, data scientists, and business analysts. By aligning data teams with business objectives and ensuring ongoing communication throughout the data lifecycle, DataOps helps organizations close the gap between data producers and consumers, ensuring that data is more actionable and aligned with business needs.

DataOps also addresses a critical challenge that DevOps helped solve: scalability. Just as DevOps practices enabled software teams to scale their operations to meet growing demand, DataOps allows data teams to scale their data pipelines in response to the rapid growth of big data. As organizations generate more and more data from various sources, the need for scalable, flexible data pipelines becomes paramount. DataOps practices, such as modular pipeline design and the use of cloud-based platforms, allow data teams to rapidly adapt to changing data volumes and sources, ensuring that data remains available, accurate, and relevant at all times.

The role of monitoring and feedback loops is another area where DataOps draws inspiration from DevOps. In the DevOps world, continuous monitoring of application performance and infrastructure health is essential for maintaining uptime and optimizing performance. Similarly, DataOps strongly emphasizes monitoring data quality, pipeline performance, and overall data health. By implementing automated feedback loops, DataOps teams can detect and resolve issues in real-time, ensuring that data remains accurate, consistent, and ready for analysis. This proactive approach to monitoring helps prevent data bottlenecks and quality issues from disrupting business operations.

While DevOps focuses on optimizing the software development lifecycle, DataOps expands this approach to include the entire data lifecycle, from data ingestion and storage to analysis and reporting. This holistic view is essential for organizations that rely on data-driven decision-making. By applying DevOps principles to data management, DataOps ensures that data is delivered continuously and reliably, empowering decision-makers to act on insights faster and with greater confidence. In many ways, DataOps represents the next logical step in the evolution of data management, aligning it more closely with the agile, automated practices that have transformed software development.

In addition to the technical parallels, the cultural impact of DevOps has also influenced the adoption of DataOps. DevOps fostered a culture of collaboration, transparency, and continuous improvement, which has proven to be just as valuable in the realm of data operations. DataOps encourages data teams to adopt a similar mindset, where experimentation is embraced, and failure is seen as an opportunity to learn and improve. This cultural shift is critical in a world where data is increasingly seen as a strategic asset, and organizations must move quickly to extract value from it.

Ultimately, the bridge between DevOps and DataOps is one of philosophy and practice. Both methodologies seek to break down silos, improve collaboration, and increase the speed and reliability of delivery—whether it's software or data. By adopting the principles of automation, continuous delivery, and cross-functional teamwork, DataOps enables organizations to build more agile, scalable, and efficient data pipelines, mirroring DevOps's success in software development. As the demand for real-time, high-quality data continues to grow, DataOps will play an increasingly important role in helping organizations harness the full potential of their data assets.

Automation and CI/CD for Data Pipelines

Automation and Continuous Integration/Continuous Delivery (CI/CD) have revolutionized software development, and their impact is now felt in the data operations world. In the context of DataOps, these practices are pivotal in ensuring that data pipelines are more efficient, reliable, and scalable. Traditionally, data pipelines were managed through a series of manual processes, often causing bottlenecks and delays. However, with the automation of CI/CD, data teams can build, test, and deploy data workflows at a much faster pace while ensuring data quality and reducing the risks of errors.

In the world of data pipelines, automation plays a crucial role in minimizing human intervention in repetitive and error-prone tasks. For example, data ingestion, transformation, validation, and even integration can be automated through a series of predefined workflows. This automation eliminates the need for manual interventions, reduces the likelihood of data errors, and speeds up data processing. In addition to improving efficiency, automation ensures that pipelines can scale up seamlessly as data volumes grow. In complex environments, where data is ingested from various sources and processed through multiple stages, automation becomes critical for managing modern data operations' increasing complexity.

The application of CI/CD principles to data pipelines is equally transformative. Just as CI/CD has enabled software development teams to deploy code changes rapidly and safely, it allows data teams to continuously improve their data pipelines without disrupting business operations. In a CI/CD pipeline, changes to data processes—whether new data sources, updates to transformation logic, or improvements in data models—are automatically tested and deployed, reducing the risk of introducing errors. This continuous integration process ensures that each change is validated in real time, allowing data teams to move faster while maintaining the integrity of the data.

A key advantage of CI/CD for data pipelines is the ability to implement automated testing. In traditional data environments, testing often occurred late in the pipeline after significant data had already been processed. This "late-stage" testing could lead to costly rework if errors were found and delays in delivering insights to business users. With CI/CD, automated testing happens early and often at each pipeline stage. This includes testing the accuracy and completeness of incoming data, validating transformation logic, and ensuring that output data meets business requirements. By catching issues early in the process, CI/CD helps to reduce downtime and provides a more consistent flow of high-quality data.

One of the most significant benefits of automation and CI/CD in data pipelines is reduced deployment time. In the past, deploying updates to a data pipeline could take days or even weeks, as teams needed to test, review, and implement changes manually. This delay meant business users had to wait longer to access critical data insights, hampering their ability to make timely decisions. By automating these processes through CI/CD pipelines, data teams can reduce the time it takes to implement and deploy changes from weeks to mere hours. This rapid deployment capability is precious in today's fast-paced business environments, where organizations must respond quickly to new opportunities or challenges.

In addition to accelerating deployment, CI/CD enhances data operations' agility and flexibility. In a world where new data sources are constantly emerging and business needs are always evolving, data pipelines must be able to adapt quickly. Through automated CI/CD workflows, data teams can easily integrate new data sources, adjust transformations, or roll out new analytics models without rebuilding entire pipelines from scratch. This agility reduces operational overhead and empowers organizations to be more responsive and innovative in their data strategies.

Moreover, automation and CI/CD improve data operations' overall governance and transparency. As data flows through automated pipelines, each step is tracked and logged, creating a detailed audit trail. This level of transparency ensures that any issues can be quickly identified and traced back to their source, making it easier to resolve problems before they affect downstream data consumers. Automated CI/CD pipelines also help enforce data governance policies, ensuring data is properly validated, secured, and compliant with regulatory standards throughout its lifecycle. This governance is critical in finance, healthcare, and retail industries, where data quality and security are paramount.

The rise of cloud-native technologies has also fueled the shift toward automation and CI/CD for data pipelines. Cloud platforms have made it easier to scale and automate data operations without the need for extensive on-premise infrastructure. Services like AWS, Google Cloud, and Microsoft Azure offer built-in tools for automating the orchestration and deployment of data workflows, allowing organizations to take full advantage of the cloud's elasticity and scalability. These cloud-based tools simplify the deployment of CI/CD pipelines, enabling teams to build and manage data pipelines in a more agile and cost-effective manner.

Finally, using automation and CI/CD in data pipelines aligns with the broader DataOps philosophy, which seeks to bring agility, collaboration, and continuous improvement to data management. DataOps is built on the principle that data should flow continuously from source to insight without unnecessary delays or bottlenecks. Automation and CI/CD are the mechanisms that enable this continuous flow, allowing organizations to deliver trusted data to business users in real time. By adopting these practices, organizations can ensure that their data operations are more efficient, resilient, scalable, and capable of supporting future growth.

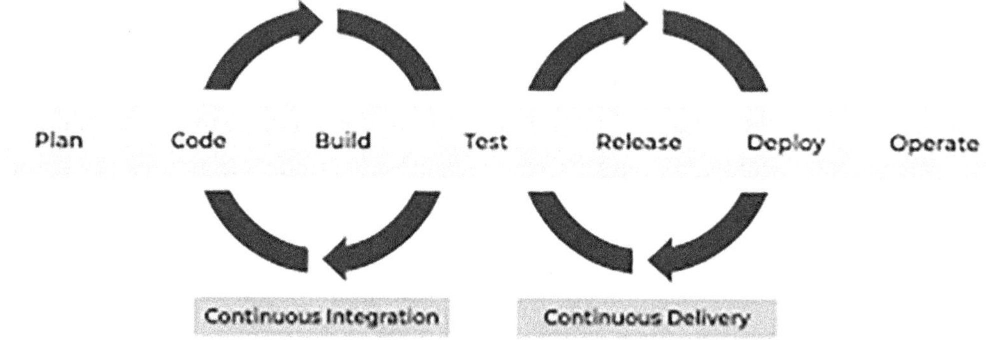

Figure 1-4. *DataOps CI/CD Process*

The Emergence of DataOps Platforms

The rise of big data and the increasing complexity of data operations have driven the demand for more sophisticated tools and platforms to manage data pipelines effectively. The emergence of DataOps platforms represents a significant leap forward in addressing the challenges associated with modern data management. These platforms are designed to streamline the entire data lifecycle, from ingestion and transformation to analysis and delivery, by providing integrated solutions that enhance collaboration, automation, and scalability.

DataOps platforms emerged in response to the limitations of traditional data management systems, which often needed help to keep pace with the rapid evolution of data sources, volumes, and business requirements. Traditional approaches relied on disparate tools and manual processes, which created inefficiencies and increased the risk of errors. DataOps platforms address these issues by offering a unified environment where data operations can be managed holistically. This integrated approach helps organizations overcome the challenges of managing complex data workflows, ensuring that data is processed quickly and accurately.

By providing features and functionality contained in a unified platform, DataOps platforms quickly outpace traditional data management tools that can sometimes be disparate and lack a cohesive ecosystem. These platforms can bring together both custom and best-of-breed tools, such as DBT and pipeline management, which can reduce technical debt, ensure the process is more efficient, and even provide developers with AI-equipped tools or features to aid in the development and deployment process. More premium tools typically contain additional functionality, such as layering in testing and pipeline observability, giving deeper insights into the DataOps process.

One of the key features of DataOps platforms is their ability to automate data pipelines. Automation is a core component of DataOps, enabling organizations to streamline repetitive tasks such as data ingestion, transformation, and quality checks. By automating these processes, DataOps platforms reduce the need for manual intervention, minimize the risk of errors, and accelerate the delivery of data insights. Automation also enhances scalability, allowing organizations to handle increasing data volumes and complexity easily. This capability is especially valuable in environments where data is generated from diverse sources and must be processed in real time.

Another significant advantage of DataOps platforms is their support for continuous integration and continuous delivery (CI/CD) of data workflows. As CI/CD transformed software development by enabling rapid and reliable code deployments, DataOps platforms apply these principles to data pipelines. With CI/CD integration, DataOps platforms facilitate the continuous testing, validation, and deployment of data processes. This approach ensures that data workflows are always up-to-date and aligned with business needs, allowing organizations to adapt to changes and maintain data accuracy and quality quickly.

DataOps platforms also play a crucial role in enhancing collaboration among data teams. In traditional data environments, communication between data engineers, analysts, and business users was often fragmented, leading to inefficiencies and misaligned objectives. DataOps platforms address this challenge by providing a centralized environment where all stakeholders can collaborate seamlessly. Features such as shared dashboards, real-time notifications, and collaborative tools enable data teams to work together more effectively, ensuring that data processes are aligned with business goals and that insights are delivered promptly.

Data governance and security are critical concerns in any data management strategy, and DataOps platforms address these needs with robust features designed to ensure compliance and protect sensitive information. DataOps platforms typically include capabilities for data lineage tracking, access controls, and auditing, which help organizations maintain visibility and control over their data assets. These features are essential for meeting regulatory requirements and ensuring data is handled securely throughout its lifecycle. Organizations can mitigate risks and build trust in their data by embedding governance and security measures into the platform.

The rise of cloud computing has also significantly influenced the development and adoption of DataOps platforms. Cloud-based DataOps platforms offer several advantages, including scalability, flexibility, and cost-effectiveness. Cloud platforms allow organizations to leverage elastic resources, scale their data operations dynamically, and avoid the high costs associated with maintaining on-premise infrastructure. Cloud-based DataOps platforms also integrate seamlessly with other cloud services, enabling organizations to build comprehensive data ecosystems that support their strategic objectives.

The need for real-time analytics and decision-making has also driven the emergence of DataOps platforms. In today's fast-paced business environment, organizations require timely access to accurate data to make informed decisions and respond to market changes. DataOps platforms are designed to facilitate real-time data processing and analysis, enabling organizations to derive insights quickly and act on them with confidence. By providing real-time visibility into data pipelines and analytics, DataOps platforms empower organizations to stay ahead of the competition and capitalize on emerging opportunities.

Furthermore, the integration of machine learning and artificial intelligence into DataOps platforms represents another significant advancement. Machine learning algorithms and AI-driven insights can be incorporated into data pipelines to enhance data quality, automate complex processes, and generate predictive analytics. DataOps platforms that support these technologies enable organizations to leverage advanced analytics capabilities, providing deeper insights and fostering data-driven innovation.

In summary, the emergence of DataOps platforms marks a transformative shift in how organizations manage and operate their data pipelines. By offering integrated solutions that automate processes, support CI/CD, enhance collaboration, and ensure governance and security, DataOps platforms address the challenges of modern data management. With the added benefits of cloud scalability, real-time analytics, and AI integration, these platforms are poised to play a central role in helping organizations navigate the complexities of big data and drive value from their data assets.

Data Quality, Governance, and Security in the DataOps Era

As organizations increasingly rely on data to drive business decisions, the importance of maintaining high standards of data quality, governance, and security has never been greater. The DataOps era has introduced new methodologies and technologies to address these critical aspects, ensuring data is accurate, trustworthy, and protected throughout its lifecycle. This focus on data quality, governance, and security is essential for maximizing the value of data while mitigating risks and ensuring compliance with regulatory requirements.

Data quality is a cornerstone of effective data operations. In the DataOps era, maintaining high data quality involves a proactive approach that integrates data validation and cleansing into the pipeline process. Automation plays a crucial role here, as DataOps platforms facilitate real-time data quality checks and automated data cleansing. Organizations can detect and address quality issues early by continuously monitoring and validating data as it flows through the pipeline, reducing the risk of inaccurate or inconsistent data affecting business decisions. This proactive approach helps ensure that data remains reliable and valuable, enabling organizations to trust their analytics and insights.

Data governance in the DataOps era has evolved to address the complexity and scale of modern data environments. Traditional data governance models, which often relied on manual processes and rigid controls, are no longer sufficient. DataOps introduces a more dynamic approach to governance, incorporating automated policies and real-time monitoring to ensure data compliance and quality. DataOps platforms provide tools for defining and enforcing data governance policies, tracking data lineage, and auditing data access and usage. This enhanced governance framework helps organizations maintain control over their data assets, ensuring that data is used appropriately and in accordance with regulatory requirements.

Data security is another critical aspect of data management that has gained increased focus in the DataOps era. With the proliferation of data and the rise of sophisticated cyber threats, protecting data from unauthorized access, breaches, and loss is paramount. DataOps platforms integrate security measures into the data pipeline, including encryption, access controls, and monitoring. By embedding security protocols directly into the data operations process, organizations can safeguard their data assets and reduce the risk of security incidents. Additionally, real-time monitoring and automated alerts help detect and respond to potential security threats promptly, enhancing overall data protection.

One of the key benefits of DataOps, as it relates to data quality, governance, and security, is its ability to automate compliance with regulatory requirements. In industries such as finance, healthcare, and retail, compliance with regulations like GDPR, HIPAA, and CCPA is critical. DataOps platforms provide built-in features for data lineage tracking, audit trails, and policy enforcement, making it easier for organizations to adhere to regulatory standards. Automation ensures that compliance measures are consistently applied across all data processes, reducing non-compliance risk and associated penalties.

Data lineage and metadata management are essential for data governance in the DataOps era. Understanding data's origin, movement, and transformation throughout its lifecycle is crucial for maintaining data quality and ensuring compliance. DataOps platforms offer robust lineage tracking capabilities, allowing organizations to visualize data flow and dependencies. This visibility helps data teams identify and address potential issues related to data quality, governance, and security. Additionally, metadata management features provide context and documentation for data assets, enhancing transparency and facilitating better decision-making.

Integrating data quality monitoring and feedback loops into DataOps platforms enhances the ability to maintain high data quality standards. Continuous monitoring of data quality metrics, such as accuracy, completeness, and consistency, allows organizations to identify and resolve issues quickly. Feedback loops enable data teams to refine and improve data processes based on quality assessments and user feedback. This iterative approach helps ensure that data quality remains high and that data operations are continually optimized to meet evolving business needs.

Collaboration and communication are crucial for adequate data quality, governance, and security. In the DataOps era, fostering a culture of collaboration between data engineers, data scientists, analysts, and business stakeholders helps ensure that data policies and standards are consistently applied. DataOps platforms facilitate this collaboration by providing shared tools, dashboards, and communication channels. By aligning data teams with business objectives and maintaining open lines of communication, organizations can better manage data quality, governance, and security, ensuring that data is accurate, secure, and used effectively.

Finally, data ethics is an emerging consideration in the DataOps era. As organizations collect and analyze vast amounts of data, ethical considerations around data use and privacy are becoming increasingly important. DataOps platforms support ethical data practices by providing features for data anonymization, consent management, and responsible data usage. By incorporating ethical considerations into data operations, organizations can build trust with stakeholders and ensure that data is used in a manner that respects individual privacy and promotes transparency.

In summary, the DataOps era has brought significant advancements in managing data quality, governance, and security. Organizations can maintain high data quality standards by leveraging automation, real-time monitoring, and integrated platforms, ensure compliance with regulatory requirements, and protect data from security threats. The focus on collaboration, transparency, and ethical data practices further enhances the ability to manage data effectively and derive valuable insights from it. As data continues to play a central role in business strategy, the importance of these aspects will only grow, making DataOps an essential component of modern data management.

The Future of DataOps: AI and ML Integration

As DataOps continues to evolve, the integration of artificial intelligence (AI) and machine learning (ML) is poised to drive transformative changes in how organizations manage their data operations. The convergence of DataOps with AI and ML technologies promises to enhance data processing, automate complex tasks, and unlock deeper insights, positioning organizations to navigate the complexities of the data-driven world better.

AI and ML integration into DataOps is set to revolutionize how data pipelines are built and managed. Traditionally, data pipelines have been designed and maintained through manual processes and predefined rules. However, data pipelines can become more intelligent and adaptive with AI and ML. Machine learning algorithms can be employed to automatically detect patterns, anomalies, and trends in data, leading to more accurate and efficient data processing. This integration allows for dynamic adjustments to data pipelines in response to changing data patterns, reducing the need for manual intervention and enhancing the overall agility of data operations.

One of the most significant benefits of incorporating AI into DataOps is enhancing data quality management. AI-driven tools can continuously monitor data for inconsistencies, errors, and quality issues, providing real-time insights and automated corrections. For instance, machine learning algorithms can be trained to identify and rectify data anomalies, perform data cleansing, and ensure that data adheres to predefined quality standards. By leveraging AI for data quality management, organizations can maintain high data accuracy and reliability levels, which is crucial for making informed business decisions.

Predictive analytics is another area where AI and ML integration into DataOps can have a profound impact. Machine learning models can analyze historical data and identify trends to predict future events or behaviors accurately. This capability enables organizations to anticipate market changes, customer needs, and operational challenges before they arise. By integrating predictive analytics into data pipelines, DataOps can provide valuable foresight and support proactive decision-making, helping organizations stay ahead of the curve and respond swiftly to emerging opportunities or risks.

Automated data pipeline optimization is an emerging trend driven by AI and ML technologies. Traditional methods of optimizing data pipelines often involve manual tuning and adjustments based on predefined metrics. With AI and ML, data pipelines can automatically optimize themselves based on real-time performance data. For example, machine learning algorithms can analyze pipeline efficiency, resource utilization, and processing times to suggest or implement optimizations that improve performance and reduce costs. This level of automation enhances the efficiency and scalability of data operations, allowing organizations to handle larger volumes of data with greater ease.

Enhanced data security through AI and ML is another critical development on the horizon. AI-driven security tools can continuously monitor data access, detect unusual patterns, and identify potential security threats in real time. Machine learning algorithms can analyze historical security incidents to predict and prevent future breaches. By integrating AI into data security measures, DataOps can provide more robust protection against data breaches and cyber threats, ensuring that sensitive information remains secure and compliant with regulatory standards.

Integrating AI and ML into DataOps supports advanced data analytics and business intelligence. AI-powered analytics tools can sift through vast amounts of data to uncover hidden insights and generate actionable recommendations. Machine learning models can enhance data visualization, automate report generation, and provide a deeper context for business decisions. By leveraging these advanced analytics capabilities, organizations can gain a more comprehensive understanding of their data, leading to more informed strategic decisions and a competitive edge in the market.

AI-driven anomaly detection is another powerful application of machine learning within DataOps. Machine learning algorithms can be trained to recognize standard patterns in data and identify deviations that may indicate issues or opportunities. This capability is invaluable for detecting fraud, operational inefficiencies, or emerging trends. By incorporating AI-driven anomaly detection into data pipelines, organizations can respond more quickly to potential issues and capitalize on new opportunities as they arise.

Looking ahead, DataOps will likely see an increasing emphasis on collaborative AI in the future. Rather than replacing human decision-makers, AI and ML will augment human capabilities by providing enhanced tools and insights. DataOps platforms will integrate AI-driven features that empower data engineers, analysts, and business users to work more effectively together. Collaborative AI will facilitate better communication, streamline workflows, and enhance the overall data management process, driving more significant innovation and efficiency in data operations.

Finally, the ethical considerations of integrating AI and ML into DataOps will be a crucial area of focus. As AI technologies become more advanced, organizations must ensure that their use of these technologies adheres to ethical standards and promotes transparency. This includes addressing issues related to data privacy, algorithmic bias, and the responsible use of AI. Organizations can build trust with stakeholders by prioritizing ethical practices and ensuring that their AI and ML integrations contribute positively to their data operations.

In summary, the integration of AI and ML into DataOps is set to transform how organizations manage their data pipelines, enhance data quality, and drive advanced analytics. By leveraging AI and ML technologies, DataOps can become more intelligent, automated, and adaptive, providing organizations with powerful tools to navigate the complexities of the data-driven world. As these technologies continue to evolve, they will play an increasingly central role in shaping the future of data operations and unlocking new opportunities for innovation and growth.

Principles of DataOps

As organizations continue to harness the power of data to drive decision-making and innovation, the need for efficient and effective data management practices has never been more critical. DataOps principles represent a set of practices and methodologies designed to enhance the management of data pipelines, ensuring that data is delivered accurately, consistently, and in a timely manner. Rooted in the best practices of DevOps, DataOps extends these principles to the data domain, focusing on improving collaboration, automation, and continuous integration within data operations.

At the heart of DataOps are several fundamental principles that collectively transform how organizations handle their data. These principles emphasize the importance of continuous integration and delivery (CI/CD) for data workflows, enabling rapid and reliable updates to data pipelines. They also highlight the critical role of collaboration and communication among data teams, ensuring all stakeholders are aligned and working towards common goals. Furthermore, DataOps principles advocate for automating and orchestrating data processes to streamline operations and reduce manual intervention, ultimately enhancing efficiency and scalability. Together, these principles form the foundation for a robust DataOps strategy that can adapt to the evolving demands of the data-driven landscape.

Continuous Integration and Continuous Delivery (CI/CD) for Data

Continuous Integration (CI) and Continuous Delivery (CD) are cornerstones of modern software development, known for their ability to streamline and accelerate the development lifecycle. When applied to data operations, these principles bring similar benefits, transforming how data pipelines are managed and deployed. CI/CD for data focuses on automating the integration, testing, and delivery of data workflows, ensuring that data updates are reliable, timely, and aligned with business needs.

Continuous Integration in data operations involves regularly merging changes to data pipelines and processes into a shared repository. This practice ensures that updates to data schemas, transformations, and integrations are consistently tested and validated. By automating the process of integrating and testing data changes, organizations can detect and address issues early, reducing the risk of errors and inconsistencies in the data. CI tools for data operations often include automated testing frameworks that validate data quality, schema integrity, and transformation logic, providing immediate feedback to data engineers and analysts.

Continuous Delivery extends the principles of CI to automate the deployment of data updates to production environments. In a DataOps framework, CD ensures that validated changes to data pipelines are seamlessly and consistently delivered to production systems without manual intervention. This process involves automating the deployment of new data workflows, transformations, and updates, which minimizes downtime and accelerates the availability of new data features and improvements. By implementing CD practices, organizations can achieve faster turnaround times for data-related changes, enhancing their ability to respond to evolving business requirements.

The benefits of CI/CD for data extend beyond operational efficiency. Organizations can improve data quality and reliability by integrating CI/CD practices into data operations. Automated testing and validation processes help ensure data pipelines function correctly and data transformations meet quality standards. This proactive approach to data management reduces the likelihood of errors and data inconsistencies, which can significantly impact business decisions. Additionally, the automation of deployment processes minimizes the risk of human error, further enhancing the reliability of data operations.

Implementing CI/CD for data also supports collaboration and transparency within data teams. Continuous integration provides a shared platform for data engineers, data scientists, and analysts to collaborate on changes and updates to data pipelines. Automated testing and deployment processes ensure that all team members work with the most current and validated versions of data workflows, fostering a more cohesive and coordinated approach to data management. This collaboration is essential for aligning data operations with business objectives and effectively supporting data initiatives.

Monitoring and feedback loops are integral components of CI/CD for data. Continuous monitoring of data pipelines and deployments provides real-time insights into the performance and health of data workflows. Automated alerts and dashboards enable teams to identify and address issues quickly, ensuring that data operations remain stable and efficient. Feedback loops allow for continuous improvement, as insights gained from monitoring can inform adjustments and refine data pipelines, driving ongoing enhancements in data quality and performance.

In summary, applying CI/CD principles to data operations offers significant efficiency, quality, and collaboration advantages. Organizations can achieve more reliable and timely data operations by automating the integration, testing, and delivery of data updates. Adopting CI/CD practices enhances the agility of data teams and ensures that data pipelines remain robust and aligned with business needs. As data environments continue to grow in complexity, CI/CD for data will play an increasingly crucial role in enabling organizations to manage and leverage their data effectively.

Collaboration and Communication in Data Teams

Effective collaboration and communication are pivotal to the success of data teams in any organization. In the context of DataOps, where data management processes are increasingly complex and interdependent, fostering a collaborative environment ensures that all team members—data engineers, data scientists, analysts, and business stakeholders—are aligned and working towards common goals. By breaking down silos and promoting open communication, data teams can enhance their efficiency, innovation, and overall effectiveness.

Collaboration in data teams often involves integrating diverse skill sets and perspectives. Data engineers focus on building and maintaining data pipelines, while data scientists analyze and interpret data to generate insights. Analysts bridge the gap between raw data and actionable business information, and stakeholders provide the context and requirements for data initiatives. Facilitating cross-functional collaboration ensures these roles work cohesively, leveraging each other's expertise to tackle complex data challenges. Tools such as collaborative platforms and shared dashboards can help streamline communication and keep everyone on the same page.

Communication is crucial in ensuring data initiatives align with business objectives and deliver value. Regular meetings, such as stand-ups or sprint reviews, allow team members to discuss progress, address challenges, and plan the next steps. Clear and transparent communication helps prevent misunderstandings and ensures all team members know project goals, deadlines, and dependencies. Documentation of data processes, decisions, and changes is also essential, as it provides a reference for team members and supports continuity in the event of personnel changes.

Shared tools and platforms enhance effective collaboration and communication. DataOps environments benefit from collaborative tools that facilitate real-time interaction and information sharing. For example, integrated project management platforms, version control systems, and data visualization tools can help teams manage workflows, track changes, and visualize data in a unified manner. These tools support collaborative efforts by providing a central hub for data-related activities and ensuring all team members can access the latest information.

Feedback loops are an essential aspect of collaboration in data teams. Regular feedback allows team members to continuously refine their processes and improve data quality. Implementing mechanisms for collecting and addressing feedback, such as review sessions and retrospectives, fosters an environment of continuous improvement. Feedback from data consumers can also provide valuable insights into how data products are used and perceived, guiding further enhancements and ensuring that data initiatives meet user needs effectively.

In summary, collaboration and communication are essential for the success of data teams operating within a DataOps framework. By fostering a collaborative environment, using shared tools, and maintaining clear communication, data teams can work more effectively together to achieve common objectives. This integrated approach enhances the efficiency and quality of data operations and ensures that data-driven initiatives deliver meaningful value to the organization.

Automation and Orchestration in Data Management

In modern data management, automation and orchestration are transformative practices that significantly enhance efficiency, consistency, and scalability. As organizations increasingly rely on complex data pipelines and large-scale data operations, automating routine tasks and orchestrating workflows become essential for managing data effectively. These practices streamline processes, reduce manual intervention, and enable teams to focus on more strategic initiatives.

Automation in data management involves using technology to perform repetitive tasks and processes without human intervention. This includes automating data ingestion, transformation, validation, and loading. For example, data pipelines can be designed to automatically extract data from various sources, apply necessary transformations, and load it into target systems. Automation reduces the risk of human error, speeds up data processing, and ensures that data workflows are executed consistently and reliably. By automating routine tasks, organizations can achieve faster data turnaround times and enhance their ability to respond to real-time business needs.

Orchestration refers to coordinating and managing automated processes and workflows across different systems and tools. In a data management context, orchestration integrates various data processing components and ensures they work together seamlessly. This might include orchestrating data pipelines, managing dependencies between different data tasks, and handling error recovery. Advanced orchestration tools provide capabilities for monitoring, scheduling, and managing complex workflows, ensuring that data operations run smoothly and that any issues are promptly addressed. Effective orchestration is crucial for maintaining the efficiency and reliability of data management processes, particularly in environments with multiple interconnected systems.

The benefits of automation and orchestration extend beyond operational efficiency. By implementing these practices, organizations can achieve improved data consistency and quality. Automated processes reduce the likelihood of discrepancies and errors arising from manual data handling. Orchestration ensures that data flows through the pipeline in a controlled and predictable manner, maintaining the integrity and accuracy of data throughout its lifecycle. This consistency is critical for generating reliable insights and supporting data-driven decision-making.

Scalability is another crucial advantage of automation and orchestration. As data volumes grow and data management needs become more complex, automated and orchestrated workflows can scale to handle increased workloads without requiring proportional increases in manual effort. This scalability is essential for organizations managing large datasets, supporting high-frequency data updates, and adapting to evolving business requirements. Automation and orchestration enable data management systems to grow in capacity and capability, aligning with the organization's expanding data needs.

Moreover, automation and orchestration support agility and flexibility in data management. By automating routine tasks and orchestrating workflows, organizations can quickly adapt to changes in data sources, business requirements, or processing needs. This flexibility allows teams to implement new data pipelines, adjust existing workflows, and incorporate new technologies with minimal disruption. As a result, organizations can remain responsive to shifting market conditions and evolving business strategies, leveraging their data more effectively to drive innovation and growth.

In summary, automation and orchestration are vital practices for modern data management. They enhance efficiency, consistency, and scalability while reducing manual effort and error. By automating routine tasks and orchestrating complex workflows, organizations can manage their data operations more effectively, supporting faster and more reliable data processing. These practices enable organizations to stay agile and responsive in a rapidly changing data landscape, ultimately driving better business outcomes and more informed decision-making.

Data Quality and Governance

In the context of DataOps, data quality and governance are integral to maintaining the integrity and reliability of data across complex and dynamic data environments. DataOps principles emphasize automation, collaboration, and continuous integration, which are crucial in enhancing data quality and implementing effective governance practices. By applying these principles, organizations can ensure that their data is accurate, consistent, and compliant with relevant standards and regulations.

Data quality is a cornerstone of effective DataOps practices. Automation in DataOps enables continuous data validation and cleansing processes, ensuring that data quality issues are identified and addressed in real time. Automated data quality checks can be integrated into data pipelines, allowing for immediate detection of anomalies, errors, or inconsistencies. This proactive approach minimizes the risk of poor-quality data affecting downstream processes and ensures that only high-quality data is used for analysis and decision-making. Automated data enrichment processes can enhance data quality by integrating additional context or correcting inaccuracies based on external sources.

Data governance in a DataOps framework focuses on embedding governance practices directly into data workflows and operations. This involves implementing automated data management policies and procedures that ensure data is handled consistently and securely throughout its lifecycle. DataOps principles advocate for continuous integration and continuous delivery (CI/CD) of data workflows, which include automated enforcement of governance policies. This ensures that data governance standards are applied uniformly, reducing the likelihood of non-compliance and errors. Organizations can maintain regulatory compliance and data integrity more effectively by incorporating governance checks into automated deployment processes.

One key benefit of integrating DataOps principles into data governance is enhanced visibility and control. DataOps practices promote the use of real-time monitoring and analytics tools that provide insights into data quality and governance metrics. This visibility allows organizations to quickly identify and address issues related to data quality and compliance. For example, monitoring tools can track data lineage, access controls, and data usage patterns, enabling teams to ensure that data governance policies are followed and to detect any deviations or potential risks.

Collaboration is another critical aspect of DataOps that impacts data quality and governance. DataOps encourages cross-functional teams to collaborate on data management and governance initiatives by fostering a collaborative environment. Data engineers, data scientists, and analysts can collaborate on defining data quality standards, establishing governance policies, and implementing automated solutions. This collaborative approach ensures that diverse perspectives are considered and that governance practices are aligned with business objectives and data needs.

Furthermore, feedback loops inherent in DataOps practices support continuous improvement in data quality and governance. Automated feedback mechanisms provide real-time insights into the effectiveness of data quality checks and governance controls. Teams can use this feedback to refine data processes, update governance policies, and address emerging issues. This iterative approach ensures that data management practices evolve in response to changing requirements and operational challenges, maintaining high data quality standards and compliance over time.

In summary, DataOps principles are pivotal in enhancing data quality and governance by leveraging automation, collaboration, and continuous integration. By integrating these principles into data management practices, organizations can ensure that data is accurate, consistent, and governed effectively. Automation streamlines quality checks and governance enforcement, while collaboration and feedback loops drive continuous improvement. Together, these practices provide a robust framework for managing data to support business objectives and regulatory compliance.

Scalability and Flexibility in Data Operations

In the rapidly evolving data management landscape, scalability and flexibility are crucial attributes for effective data operations. As organizations face increasing data volumes, diverse data sources, and dynamic business requirements, the ability to scale data operations and adapt to changes becomes essential. DataOps principles provide a framework for achieving these goals, ensuring that data systems can efficiently handle growth and adapt to shifting needs.

Scalability in data operations refers to the capability of data systems and processes to handle growing data volumes and workloads without compromising performance. DataOps principles, such as automation and continuous integration, support scalability by enabling organizations to manage and scale their data pipelines efficiently. For example, automated scaling features in cloud-based data platforms allow organizations to adjust resources based on data processing demands. This ensures that data operations can accommodate increased data loads and user queries without manual intervention, maintaining optimal performance and responsiveness.

Flexibility is equally important in data operations, allowing organizations to adapt to changing business requirements and data environments. DataOps practices emphasize using modular and configurable data pipelines, which can be easily adjusted or extended to meet new needs. By designing data workflows with flexibility in mind, organizations can incorporate new data sources, modify existing processes, and integrate emerging technologies with minimal disruption. This adaptability is crucial for staying competitive in a data-driven market and responding quickly to evolving business conditions.

One of DataOps' key advantages in achieving scalability and flexibility is its focus on automated testing and deployment. Automated testing ensures that changes to data pipelines and processes are thoroughly validated before deployment, reducing the risk of errors and ensuring that scaling efforts do not introduce new issues. Continuous deployment practices enable organizations to release updates and enhancements to data systems rapidly and reliably. This agility allows organizations to scale their operations and adapt to changes more efficiently, supporting business growth and innovation.

Monitoring and analytics are also vital for maintaining scalability and flexibility in data operations. DataOps practices incorporate real-time monitoring and performance analytics to provide insights into the health and efficiency of data systems. Organizations can continuously monitor data pipelines and resource usage to identify potential bottlenecks, performance issues, or capacity constraints before they impact operations. This proactive approach enables timely adjustments and optimizations, ensuring data systems remain scalable and flexible in response to changing demands.

Furthermore, collaboration and communication within data teams contribute to scalability and flexibility. DataOps promotes a collaborative environment where data engineers, data scientists, and business stakeholders work together to design and manage data systems. This collaboration facilitates a shared understanding of scalability requirements and helps align data operations with business goals. Effective communication ensures that all team members are informed of changes, updates, and scaling strategies, supporting a coordinated approach to managing data operations.

In summary, scalability and flexibility are essential for managing data operations effectively in a dynamic environment. DataOps principles, including automation, continuous integration, and modular design, support these attributes by enabling organizations to scale their data systems and adapt to evolving needs. Automated testing and deployment, real-time monitoring, and collaborative practices further enhance the ability to efficiently manage growing data volumes and changing requirements. By leveraging these principles, organizations can maintain high performance and agility in their data operations, driving success in a data-driven world.

Benefits of Adopting DataOps

When I talk to my clients about DataOps and its benefits, I usually focus on minimizing the impact on production systems. I firmly believe that data is the lifeblood of any organization, and any loss of data or downtime to that data can significantly impact how a business can perform its daily operations. If you continue that line of thought, you'll quickly see how productivity and profits start falling, and costs can increase. Although this only scratches the surface of how DataOps can benefit your team or organization, it is typically the critical element that resonates with relevant stakeholders. What I mean by "impacts to production systems" is my way of asking, "How do you deploy new changes to your data environment without causing a loss of data or experiencing an outage?"

Building on this, I will assume that you are not building changes directly in your production database. If this is you, put this book down and pump the brakes on your team as much as possible. If you already have at least one non-production environment where you can test your changes before you deploy them, I will say you are already on the right track to setting up a solid DataOps process. However, one critical element is probably missing from your approach. How do you ensure that what you did in your non-production environment is repeatable, exactly, in your production environment? Suppose you're saving a script that you manually execute in production after you do a functional check in a lower environment; let's call it dev. In that case, you're still at risk of causing a loss of data or breaking production.

The DataOps Process

DataOps minimizes this risk by allowing you to promote changes from a non-production environment to your production system. A good DataOps process will have multiple layers of non-production systems that code touches before its production even gets a glimpse of the change. We call this process "promotion," and these promotions run through a "pipeline" (Figure 1-5). Moreover, DataOps promotes your code and handles the ongoing integration of your data. As I have said before, your data is continually growing and evolving. Whether you get data daily in your data environment or, like with many IoT-based companies, that data comes in via a real-time schedule, all of that data can flow through a pipeline. The process of data flowing through a pipeline allows you to set up gatekeepers on your data, or quality checks, to ensure the data is true and accurate before it creates a dirty production environment.

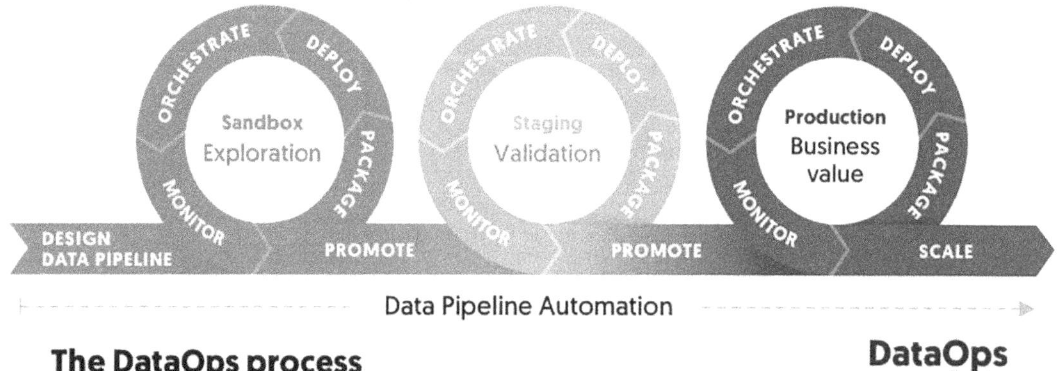

Figure 1-5. *The DataOps Process*

In Figure 1-5, you can see this pipeline process in a visual representation. This process applies specifically to data engineering, and you might use a more straightforward pipeline visualization to represent how data flows through a pipeline. In my Figure 1-5 example, all development starts in some type of sandbox. Perhaps you call this "dev," and all users work within it, or every developer has their sandbox, or you even create sandboxes on demand for each discrete element of work you are performing (feature branches). The development pipeline handles your promotion to the next pipeline stage, in this case, a staging environment. Staging should never see direct development, and it ensures that what you develop in a sandbox can be repeated without user input or intervention. After performing a quality check in staging, you can trigger the pipeline to promote to production. When you make the final promotion to production, you have much higher confidence that the changes will perform as expected and not adversely affect your production system. While this example is sound foundationally, I encourage my clients to consider two staging environments between their sandbox(es) and production. The first of these two staging environments is how your team validates their expectations. The second of the two staging environments, what I would call a UAT environment, lets your end users (potentially even your integrated systems) validate that everything looks as expected.

DataOps Pipelines for Data

Understanding the core concepts of DataOps will help you rapidly establish a foundation you can build upon for a robust DataOps solution in your organization. To get a complete picture, however, it is essential to remember that we look at data development holistically. Data engineering, think of this like a software engineer, is one piece of the puzzle. They are responsible for building the house where the data lives. The second element is the actual data, which we will now look at.

Data enters our data centers in various ways, and not all of those methods require you to think about DataOps pipelines. If, for example, you have data that comes in via a web service that is writing directly to your core tables, a pipeline is likely unnecessary. Instead, you should consider how data moves *inside* the data center. I will use a basic example of how data moves inside a data center, such as Snowflake, that follows a Medallion design pattern.

Assume that you have data coming into a series of tables that you categorize as a data lake. This data is entirely untouched as it comes in from your source system but ultimately will require some "massaging of the data" as it makes its way through

Snowflake and ultimately to a place where your business users or systems can consume it. In this setup, you might have a data lake for the raw data, a data warehouse for the transformed and enriched data, and a data mart where you perform metrics and prepare the data for consumption by the business. Data moving through all of the stages in Snowflake, starting from the data lake, is an ideal candidate for DataOps pipelines.

You can utilize existing technologies like DBT to help with this movement, streamline code, and leverage repeatable macros (Figure 1-6). Tools like DBT are instrumental in creating repeatable, maintainable, and reliable code that helps ensure you process your data the same way every time. Inherently, tools like DBT add a layer of testing and documentation, both core tenets of the DataOps methodology.

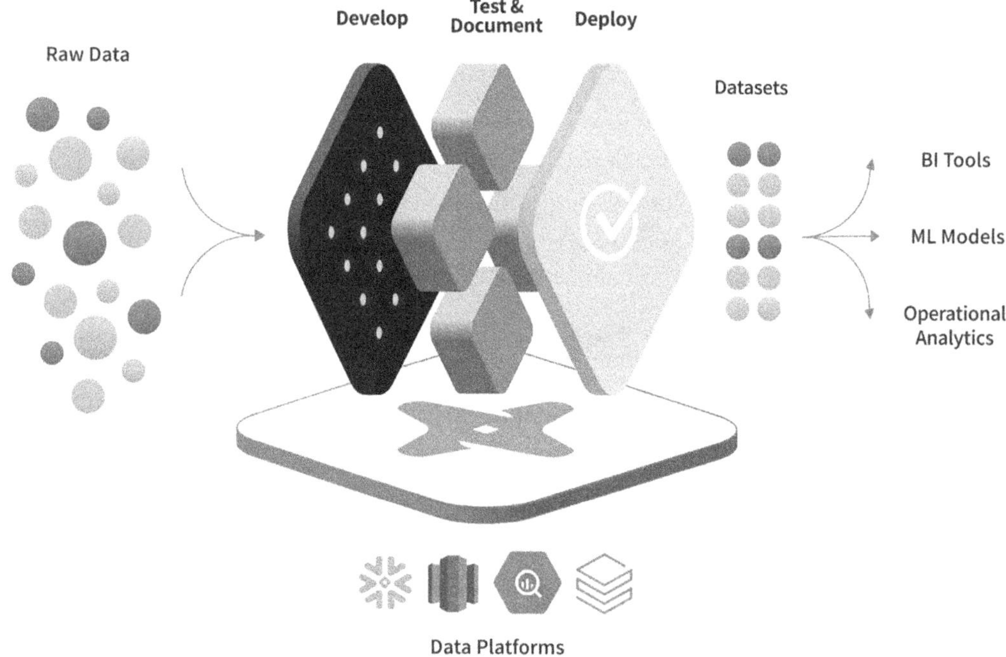

Figure 1-6. *DBT Data Flow*

In Figure 1-6, you can see a visual representation of how DBT takes raw data and passes it through a structured process. The result is highly detailed datasets that end users, ML models, or other business tools can consume. The primary goal that DataOps aims to achieve for your data is a thoroughly tested and trusted output of datasets that your business can rely on to make critical decisions.

Conclusion

As data management has evolved, the principles of DataOps have played an essential role in shaping how organizations handle, process, and derive value from their data. DataOps bridges the gap between traditional database engineering and modern data-driven decision-making by promoting automation, agility, and collaboration. These foundational principles directly affect machine learning and AI, where data quality, consistency, and governance are paramount. Without a strong data pipeline, even the most sophisticated AI models will struggle to deliver meaningful insights or maintain accuracy over time.

The transition from traditional data models to those optimized for AI underscores the importance of a well-orchestrated data lifecycle. Data preparation, feature engineering, and pipeline automation are critical in ensuring machine learning models perform efficiently and scale effectively. Much like DataOps principles advocate for continuous integration, testing, and deployment, AI-driven workflows require iterative refinement, monitoring, and governance. By integrating DataOps methodologies into AI and ML processes, organizations can create a seamless, high-quality pipeline that ensures reliable, real-time data for their models.

As AI and machine learning become more pervasive, organizations must rethink their data strategies to accommodate the dynamic nature of these technologies. This means moving beyond static data models toward event-driven, graph-based, and real-time architectures supporting AI-driven applications. The combination of structured and unstructured data, the need for feature stores, and the automation of data transformation all require a holistic approach that aligns with the core principles of DataOps. Emphasizing reproducibility, data lineage, and governance ensures that AI models remain trustworthy and unbiased in production environments.

By applying the lessons from DataOps to AI and machine learning, organizations can create more resilient, scalable, and efficient data architectures. This chapter explores how data modeling practices must evolve to support AI-driven initiatives, highlighting the importance of data preparation, feature engineering, and automation. As we continue our journey into modern data management, the next frontier will be integrating these concepts with real-time and event-driven architectures, further pushing the boundaries of what is possible in data-driven innovation.

CHAPTER 2

Pillars of True DataOps

In the realm of DataOps, the #TrueDataOps Community has identified a set of foundational principles that are essential for achieving effective and efficient data operations. These principles, often referred to as the "Pillars of True DataOps," represent core practices that drive successful data management and operational excellence. This chapter explores these seven pivotal pillars: ELT (Extract, Load, Transform) and the principles underlying it, agility through CI/CD (Continuous Integration and Continuous Delivery), component design and maintainability, environment management, governance, security, and change control, automated testing and monitoring, and the importance of collaboration and self-service (Figure 2-1). These pillars contribute to a robust DataOps framework, ensuring that data operations are scalable, reliable, and aligned with organizational goals.

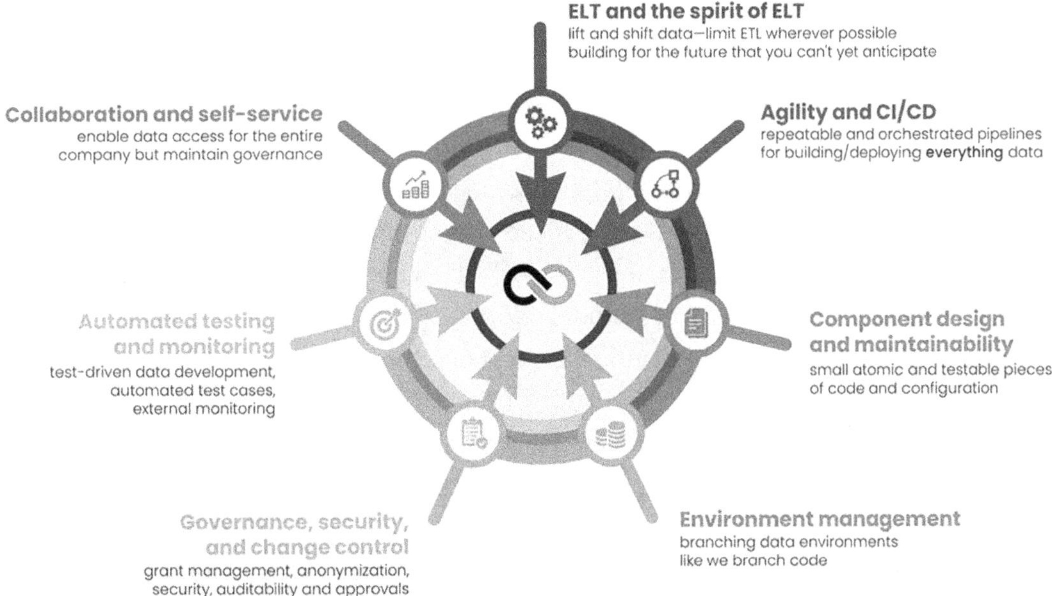

***Figure 2-1.** Seven Pillars of #TrueDataOps (Credit: truedataops.org)*

R. L. Steelman, Jr., *Mastering Snowflake DataOps with DataOps.live*, https://doi.org/10.1007/979-8-8688-1754-0_2

Pillar 1: ELT (and the Spirit of ELT)

The ELT (Extract, Load, Transform) model represents a shift from the traditional ETL (Extract, Transform, Load) approach, which is particularly well-suited for modern data lake implementations. Unlike ETL, which processes and transforms data before it enters the data repository, ELT allows data to be loaded into the system in its raw form. This strategy facilitates faster data ingestion and provides flexibility in handling data, as transformations can be applied later in the data lifecycle. By deferring data transformation, organizations avoid the constraints of pre-defining schemas and queries, leading to a more adaptable and efficient data management process.

The fundamental advantage of ELT is its ability to preserve the original state of the data. This "lift and shift" approach ensures that no data is lost during the loading process, which is particularly valuable when the full scope of data usage is not yet determined. By maintaining data in its raw form, ELT maximizes its utility for future analysis, allowing organizations to avoid premature data reduction or loss.

A crucial benefit of the ELT approach is its alignment with data governance principles and auditability. Keeping data in its original format makes it easier to track its lineage and verify its accuracy. This transparency simplifies compliance with governance standards and facilitates auditing processes. In cases where sensitive information must be protected, ELT's EtLT (Extract, transform, Load, Transform) variant can be used. This approach involves performing necessary transformations, such as data anonymization or encryption, before loading the data into the system, ensuring that sensitive information is appropriately handled.

Data management often involves challenges related to data loss, particularly when historical data is overwritten or replaced. For example, when reloading a table containing employee information, there is a risk of losing valuable historical context if the data is not preserved. The ELT model addresses this issue by maintaining data integrity and history, thereby preventing the loss of critical information during data operations.

The "Spirit of ELT" extends beyond the technical process of loading and transforming data. It embodies a data preservation philosophy, advocating for retaining all potentially valuable information. This principle emphasizes the importance of avoiding actions that might eliminate valuable data or limit future analytical possibilities. While implementing comprehensive data preservation solutions can be complex and costly, making these solutions more accessible and efficient is vital to realizing the full benefits of ELT.

In summary, ELT and its associated principles provide a robust framework for managing data in modern environments. Organizations can achieve greater flexibility and efficiency in their data operations by prioritizing raw data preservation and deferring transformations. The ELT model supports strong data governance and compliance, while the Spirit of ELT ensures that valuable data remains accessible for future use. Embracing these practices enhances data management and positions organizations to leverage their data more effectively.

Pillar 2: Continuous Integration/Continuous Delivery (CI/CD)

Continuous Integration (CI) and Continuous Delivery (CD) are well-established practices in software development, where code changes are frequently integrated into a central repository, built, tested, and prepared for deployment. In the realm of DataOps, these principles are equally vital but adapted to manage data pipelines and workflows. CI/CD in the context of data operations involves continuously integrating new data processes and transformations into a unified system, ensuring that changes are promptly tested and deployed without disrupting ongoing operations.

The essence of CI/CD for data operations lies in integrating data changes with the same frequency and rigor as code updates. This involves employing a robust revision control system to manage data transformations, logic, and pipeline configurations. By maintaining a central repository for all artifacts related to data workflows, organizations can ensure that every change is tracked, tested, and deployable at any time. This practice minimizes the risk of introducing errors and ensures that issues are identified and addressed immediately, fostering a more agile and resilient data management environment.

In DataOps, treating data logic and transformations as a cohesive code base is crucial. This approach mirrors software development principles, where each data pipeline is managed with stringent governance and version control. By applying these principles, organizations can achieve a consistent and controlled environment for data operations, enabling iterative improvements and rapid deployment of data changes. This method not only streamlines the development process but also enhances data workflows' overall quality and reliability.

A significant advantage of adopting CI/CD in data operations is the ability to support continuous prototyping and testing. Unlike traditional development models that require a fixed set of requirements before starting work, DataOps allows for the creation of branches that include configurations, code, and data. These branches serve as isolated environments where new data processes can be developed and refined without impacting the primary system. This flexibility enables teams to experiment and iterate on data transformations and integrations, aligning closely with stakeholder requirements and adapting to evolving needs without compromising data quality.

Moreover, DataOps aims to simplify and even automate data lifecycle management. Organizations can reduce the complexity of managing data environments by leveraging cloud-based platforms like Snowflake, which inherently support efficient data management and storage solutions. Snowflake's integration with cloud storage technologies such as AWS S3, Google Cloud Storage, and Azure Blob Storage eliminates many traditional challenges associated with data lifecycle management. These platforms provide automatic data compression and scalable storage options, streamlining environment management and making data lifecycle processes more seamless and cost-effective.

In summary, the principles of CI/CD bring agility to data operations by enabling frequent integration, testing, and deployment of data processes. By managing data transformations with the same precision as software code and adopting cloud-based solutions that automate data management, organizations can enhance their ability to respond quickly to changing requirements and maintain high-quality data operations. This approach supports continuous improvement and aligns data practices with the evolving landscape of data technology.

Pillar 3: Component Design and Maintainability

The advent of cloud computing has revolutionized how organizations approach their computing resources. With virtually unlimited computational power at their disposal, the focus of #TrueDataOps has shifted from merely optimizing CPU usage to enhancing developer productivity. By leveraging the scalability and flexibility of cloud technologies, teams can concentrate on creating efficient, maintainable code rather than obsessing over the efficiency of every CPU cycle. This shift allows developers to spend more time iterating on features and accommodating stakeholder feedback without being bogged down by premature optimization.

In software engineering, maintainability refers to how easily a system or component can be maintained over its lifecycle to minimize downtime and ensure ongoing functionality. The principles for designing maintainable code have been developed through extensive experience and practice. Well-established guidelines, such as those outlined by Eric Raymond in *The Art of Unix Programming*, remain relevant today. Raymond's principles emphasize the importance of creating modular, readable, and simple programs, advocating practices for building small, robust components and avoiding unnecessary complexity.

Raymond's principles suggest that maintainability can be significantly improved by adopting several key practices. For instance, modular programming allows you to easily make updates and modifications while writing readable and transparent code and facilitates debugging and collaboration. The idea of using composition over inheritance and prototyping before finalizing the design ensures that the code remains flexible and adaptable. These practices are designed to enhance code robustness and extendibility, ensuring that changes and enhancements can be made with minimal impact on the overall system.

The #TrueDataOps philosophy aligns with these timeless principles by emphasizing small, reusable components. By focusing on modular design and preferring configuration over custom code where possible, DataOps practices streamline the development process. Low-code solutions are utilized to reduce the amount of hand-written code further, making it easier to implement, test, and refine individual components. This approach accelerates development and simplifies the process of updating or replacing components without disrupting the end-user experience.

Ultimately, the goal of adopting a component-based approach within #TrueDataOps is to enhance the maintainability and scalability of data systems. Organizations can prototype, develop, and optimize their data workflows more efficiently by breaking down complex systems into manageable atomic pieces. This results in a more agile development process and a more resilient data architecture that can adapt to evolving business needs while minimizing user impact.

Focusing on component design and maintainability allows organizations to take full advantage of cloud computing resources while ensuring that their data systems remain robust and adaptable. By adhering to proven principles and embracing a modular, low-code approach, #TrueDataOps supports the creation of scalable, maintainable solutions that enhance developer productivity and align with modern data management practices.

Pillar 4: Environment Management

Environment management is one of the most intricate challenges within #TrueDataOps. Unlike traditional web development, where teams have successfully established multiple long-lived environments—such as production, quality assurance, and development—data environments have lagged behind in this evolution. In web development, the ability to create dynamic, purpose-specific environments, often referred to as "feature branches," allows engineers to conduct comprehensive integration tests efficiently. Technologies like Kubernetes enable the rapid deployment of new environments, ensuring that development processes remain agile and responsive.

In contrast, many organizations involved in data operations typically maintain only two or three manually configured long-lived environments. This manual approach often leads to discrepancies between environments, as they can quickly diverge due to the complexities of updating and maintaining them. The time spent managing these differences can detract from more strategic initiatives, making efficient environment management a critical area for improvement in the DataOps landscape.

Effective environment management in #TrueDataOps requires automation at every level. All environments should be provisioned, modified, and removed automatically to streamline operations. Additionally, a vital aspect of this automation is the ability to replicate production environments quickly. This often necessitates duplicating large volumes of data, which can be a significant hurdle for many organizations.

The Snowflake data platform exemplifies an innovative solution in this regard, offering features like Zero Copy Cloning that facilitate the creation and management of environments with minimal overhead. Such capabilities can revolutionize data analytics by allowing teams to quickly spin up environments that mirror production settings without data duplication and management constraints. As more cloud data platforms adopt similar features, they will likely transform how organizations approach environment management, making it as agile and efficient as contemporary web development practices.

When implemented correctly, automated environment management can significantly enhance the efficiency of a data platform's development process. It reduces operational costs by minimizing manual efforts to create, maintain, and verify environments. Organizations can focus their resources on more value-added activities by streamlining these processes, fostering a more responsive and effective data management ecosystem.

Pillar 5: Governance and Change Control

In #TrueDataOps, embedding governance and privacy into the design of data operations is not merely an afterthought but a foundational principle. A robust #TrueDataOps platform must incorporate strong governance frameworks and privacy measures from the outset rather than attempting to retrofit these critical elements later in the process. This proactive approach is essential when organizations face substantial financial penalties for mishandling data. By integrating governance and privacy into the core of data operations, organizations can mitigate risks and ensure compliance with ever-evolving regulatory landscapes.

Under the #TrueDataOps model, every modification to the data pipeline is subject to rigorous automated testing, which helps to identify errors and potential security vulnerabilities before they escalate. This commitment to quality assurance is complemented by a policy requiring multiple levels of oversight—ensuring that all changes undergo a thorough review process involving at least two sets of eyes. This dual-review mechanism fosters accountability and helps maintain the integrity of data operations, providing an added layer of security against mistakes and omissions.

Establishing a definitive source of truth is also critical in #TrueDataOps. This source serves as the benchmark for measuring data quality and accuracy. In software development, the source of truth is typically the code repository, which provides the foundation for the application's functionality. Similarly, in the context of #TrueDataOps, the code repository supporting data processes and components is considered the authoritative source for validating and auditing data operations. This alignment ensures all stakeholders have a consistent reference point, enhancing collaboration and clarity across the data lifecycle.

Implementing automated audit trails further bolsters governance in a #TrueDataOps environment. These comprehensive logs track every change, test, and approval, creating a transparent record that can be referenced indefinitely. Organizations can demonstrate compliance and accountability by maintaining detailed documentation of how data is accessed and utilized. This auditability satisfies regulatory requirements and builds trust with stakeholders, reinforcing the organization's commitment to ethical data management practices.

Adopting Governance and Privacy by Design within #TrueDataOps is essential for fostering a secure and compliant data environment. By prioritizing these principles from the start, organizations can confidently navigate the complexities of data management, ensuring that their operations remain efficient and responsible.

Pillar 6: Automated Data Testing and Monitoring

Traditionally, organizations have faced a paradox: the more they invest in their data platforms, the more complex and costly future development becomes. Increased complexity and platform sprawl, combined with an expanding codebase, complicate testing and resource allocation. As companies scale their data operations, they often struggle to manage the associated overhead, making it imperative to seek solutions that streamline processes and reduce costs.

The prospect of comprehensive testing is daunting for data teams already overwhelmed by stakeholder demands. Automated data testing emerges as a crucial solution to alleviate this burden. By proactively identifying issues before they escalate, automated testing frees up team resources and enhances the overall reliability of data operations. In an agile development environment, where the expectation is rapid deployment, balancing optimism—believing everything will work perfectly on the first try—against a solid testing framework becomes essential.

Leading organizations like Amazon, Netflix, and Etsy exemplify the power of automated testing in enabling quick and reliable deployments, sometimes rolling out updates every few seconds. Their success in managing millions of lines of code is attributable primarily to their robust automated testing practices, which have driven significant efficiency improvements in software development. The principle is straightforward: to confidently release updates, every system part must be thoroughly tested. As systems expand, the workload increases, making automation the only feasible solution for maintaining test coverage while accelerating deployment cycles.

However, automated testing alone is insufficient. It is vital to implement automated monitoring to address the reality that, despite best efforts, some issues will inevitably slip through the cracks. While automated testing seeks to prevent problems before they reach production, monitoring ensures that data quality or system performance discrepancies are quickly identified and addressed. In a #TrueDataOps framework, monitoring extends beyond simple availability checks; it includes data quality assessments and the integrity of analytics processes. This holistic approach safeguards against unexpected variations and provides valuable insights that can further optimize performance.

Furthermore, the definition of data availability has evolved within the #TrueDataOps context. It now encompasses the ability to run queries and the capacity to deliver valid and actionable insights for decision-making. A platform might boast impressive uptime statistics, but its effectiveness is undermined if it cannot provide reliable data. Broken data pipelines can compromise analytics, leading to flawed decision-making. These failures may go unnoticed without automated testing and monitoring, highlighting the critical need for robust mechanisms to maintain data integrity in today's fast-paced data environments.

Pillar 7: Collaboration and Self-Service

Collaboration and self-service are integral components of a successful #TrueDataOps framework, manifesting in two primary dimensions. First, at the operational level, diverse teams—including data engineers, analysts, data scientists, and machine learning specialists—contribute to developing and maintaining data products. Each team brings its own specialized tools and methodologies, which can lead to silos that hinder effective communication and coordination. In a traditional setup, these teams often operate independently, resulting in unpredictable product quality and challenges in aligning efforts across the data pipeline.

The #TrueDataOps model aims to streamline this collaborative process, transforming the data production lifecycle into a well-orchestrated operation akin to an efficient manufacturing line. Fostering an environment where various teams can interact seamlessly emphasizes the importance of a heterogeneous tooling ecosystem that facilitates efficient collaboration. This orchestration enhances the predictability and quality of the data products delivered and ensures that all stakeholders can access the insights they need on time.

On a broader scale, #TrueDataOps also addresses the challenges faced by business users seeking to leverage data for informed decision-making. While the ambition to create a data-driven culture is shared across organizations, the practical execution often falls short. Many business users struggle to navigate the complex landscape of available data, leading to a reliance on restricted access governed by departmental silos. To truly democratize data access and enable exploration, a robust data catalog must be implemented, allowing users to discover and understand the datasets at their disposal quickly. Additionally, creating anonymized versions of sensitive data can facilitate broader sharing without compromising privacy, and establishing lightweight processes for user-generated data ensures that valuable insights can emerge organically, fostering a more data-driven organizational culture.

Conclusion

The principles of True DataOps provide a structured, automation-first framework for managing modern data operations efficiently and reliably. By emphasizing ELT, agility, governance, continuous testing, collaboration, and self-service, these pillars create an ecosystem where data is managed and actively optimized for performance, security, and scalability. Through automation, version control, and CI/CD practices, teams can ensure that data remains a high-value asset rather than a bottleneck in innovation. Organizations that embrace these principles can achieve greater consistency, transparency, and efficiency in their data workflows.

As we move forward, it's crucial to recognize that DataOps is not just about processes and technologies—it is a cultural shift that demands collaboration between data engineers, analysts, and business stakeholders. The goal is to break down silos, enhance visibility, and ensure that data flows seamlessly across an organization in a controlled yet flexible manner. The ability to rapidly iterate, deploy changes safely, and maintain governance ensures that data-driven decision-making remains agile and trustworthy. This alignment between technical execution and strategic objectives makes True DataOps a transformative approach.

While the core principles of DataOps provide the foundation for managing structured data at scale, integrating AI and machine learning into this ecosystem presents new challenges and opportunities. Traditional data workflows were built with structured analytics, but AI-driven models require continuous data ingestion, feature engineering, and real-time adaptability. This introduces the need for specialized approaches, such as MLOps, which extend the automation and governance principles of DataOps into the world of machine learning. By understanding the relationship between these disciplines, organizations can ensure that their AI initiatives are built on a strong data foundation.

The next chapter will explore MLOps and how it bridges the gap between machine learning development and production deployment. Just as DataOps has revolutionized data management, MLOps applies similar principles to ensure AI models remain scalable, reproducible, and governed throughout their lifecycle. As organizations strive to harness the power of AI, understanding the interplay between DataOps and MLOps will be critical in building sustainable, data-driven solutions.

CHAPTER 3

MLOps

In today's data-driven world, machine learning (ML) has evolved from a research-focused field into a critical component of modern business operations. However, the challenges of deploying, scaling, and maintaining ML models in production have grown exponentially. This is where MLOps—a portmanteau of "Machine Learning" and "Operations"—comes into play.

MLOps combines the principles of DevOps with the unique needs of machine learning workflows. It provides a structured approach to streamline the end-to-end lifecycle of ML systems, ensuring efficiency, reproducibility, and scalability. This chapter will delve into the core principles of MLOps, explore the tools and technologies that enable it, and provide actionable insights into building robust MLOps pipelines. Whether you're a data scientist, ML engineer, or business leader, understanding MLOps is essential for leveraging the full potential of AI in your organization.

What Is MLOps?

MLOps, or Machine Learning Operations, is a discipline that applies the principles of DevOps to the machine learning lifecycle (Figure 3-1). MLOps is designed to streamline ML models' development, deployment, and maintenance, ensuring they operate efficiently, reliably, and at scale in production environments. While DevOps focuses on automating and improving the development and deployment of software, MLOps extends these practices to encompass the unique challenges of ML workflows, such as managing data pipelines, retraining models, and monitoring their performance in real time.

R. L. Steelman, Jr., *Mastering Snowflake DataOps with DataOps.live*, https://doi.org/10.1007/979-8-8688-1754-0_3

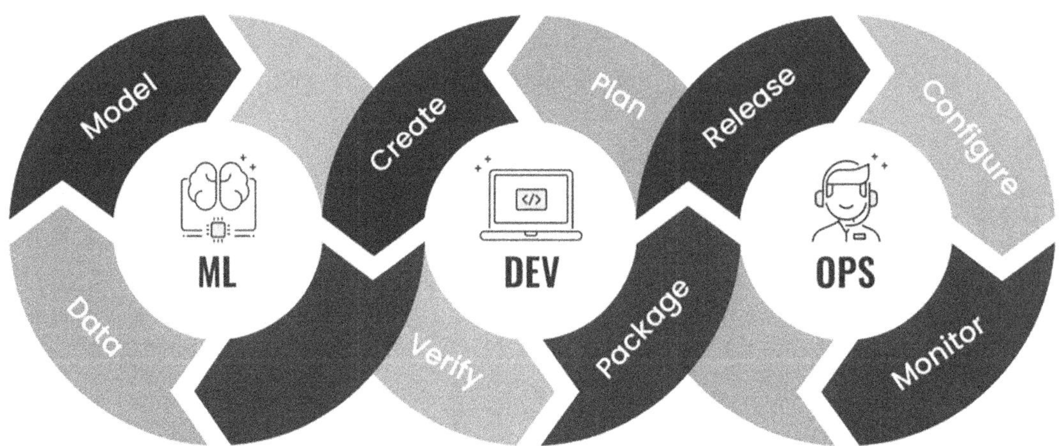

Figure 3-1. *MLOps Loop*

The core premise of MLOps is to bridge the gap between data science and production engineering. Traditionally, data scientists work in isolated environments, experimenting with different datasets and algorithms to create ML models. Once a model is ready, deploying it into a production environment often involves manual handoffs to engineers, leading to delays, errors, and misaligned expectations. MLOps eliminates this bottleneck by providing an integrated framework where data scientists, engineers, and IT teams can collaborate seamlessly.

At its heart, MLOps emphasizes automation and reproducibility. Automating data preparation, model training, and deployment tasks allows teams to iterate quickly and consistently. Reproducibility ensures that models can be retrained and deployed under the same conditions, activities that are essential for debugging, auditing, and compliance. By formalizing these workflows, MLOps reduces the risk of introducing errors during transitions between development and production.

Another key aspect of MLOps is continuous integration and continuous deployment (CI/CD) for machine learning. In the context of software development, CI/CD enables rapid iteration and frequent releases. For ML, CI/CD ensures that models are regularly updated with the latest data, tested rigorously, and redeployed efficiently. The goals of the CI/CD processes are particularly critical in dynamic environments where data patterns evolve over time, leading to phenomena like model drift, where model accuracy degrades as the underlying data changes.

MLOps also addresses the complexity of managing the interdependencies between data, code, and trained models. Unlike traditional software, ML systems rely on ever-changing datasets that influence the system's behavior. This necessitates robust version control not only for code but also for datasets and model artifacts. MLOps frameworks track these components, enabling teams to reproduce experiments, debug issues, and roll back to previous versions when necessary.

Monitoring and maintenance are integral to MLOps. Once an ML model is deployed, tracking its performance and detecting anomalies such as data drift, bias, or unexpected behaviors is essential. Monitoring tools provide metrics that help teams identify when a model needs retraining or requires intervention to prevent business disruptions. This proactive approach minimizes downtime and ensures the system continues to deliver accurate and actionable insights.

Furthermore, MLOps plays a vital role in governance and compliance. With growing concerns over data privacy, security, and ethical AI, organizations must demonstrate accountability in their ML practices. MLOps frameworks provide audit trails, documenting every stage of the ML lifecycle, from data acquisition to model deployment. These records are crucial for satisfying regulatory requirements and building trust with stakeholders.

Ultimately, MLOps represents a paradigm shift in how organizations approach machine learning. It transforms ML from a one-off, experimental activity into a scalable, reliable, and integral part of business operations. By embracing MLOps, organizations can harness the full potential of their data and deliver ML-driven solutions with speed, consistency, and confidence.

The MLOps Lifecycle

The MLOps lifecycle represents the end-to-end process of developing, deploying, and maintaining machine learning models in production. It encompasses a series of interconnected stages, each addressing specific challenges in ensuring that machine learning systems are scalable, reliable, and continuously improving. By standardizing these stages, the MLOps lifecycle enables organizations to operationalize ML workflows efficiently, bridging the gap between data science experimentation and production readiness.

The lifecycle includes key phases such as data collection and preprocessing, model development, training, validation, deployment, monitoring, and retraining. Each stage must be carefully managed to prevent model drift, maintain performance, and ensure compliance with governance and regulatory requirements. Unlike traditional software development, where code is relatively static, ML models evolve as they encounter new data, making automation and continuous integration essential. MLOps introduces best practices like versioning, CI/CD for ML models, and automated monitoring to streamline this ongoing evolution.

This structured approach to machine learning deployment helps solve various operational challenges. For instance, models trained on historical data may degrade in accuracy as real-world data distributions shift over time. Automated model monitoring and retraining pipelines can detect these changes and trigger necessary updates, reducing the risk of outdated or biased predictions. Additionally, MLOps frameworks support collaboration between data scientists, engineers, and business teams, ensuring that model development aligns with business needs while maintaining transparency and reproducibility.

By bridging the gap between ML development and production, MLOps streamlines model lifecycle management in healthcare, finance, and e-commerce industries. It enables organizations to deploy AI-powered solutions confidently, minimizing downtime, improving decision-making, and reducing risks associated with unreliable models. Whether optimizing fraud detection in banking, improving patient diagnostics in healthcare, or personalizing recommendations in retail, the MLOps lifecycle ensures that AI models remain robust, scalable, and adaptable in dynamic environments.

1. Problem Definition and Business Understanding

The MLOps lifecycle begins with a clear understanding of the problem to be solved and its business context. This stage involves identifying specific objectives, aligning them with organizational goals, and defining success criteria. Collaborating with stakeholders ensures the solution will address real-world needs, such as improving customer experiences or optimizing operations. This phase also sets expectations for model performance, timelines, and resource allocation.

2. Data Collection and Preparation

Data is the foundation of any ML project, and its quality significantly impacts the model's performance. The data collection phase involves sourcing relevant datasets from various platforms, APIs, or databases. Once collected, the data must be cleaned, transformed, and enriched to ensure it is ready for analysis. Tasks such as handling missing values, normalizing features, and identifying outliers are part of data preparation. Automation is crucial here, as it standardizes data pipelines and ensures reproducibility.

3. Feature Engineering and Selection

Feature engineering involves creating new variables (features) that better represent the underlying patterns in the data. This step often requires domain expertise to derive meaningful insights from raw data. On the other hand, feature selection involves choosing the most relevant features to improve model performance and reduce complexity. Practical feature engineering and selection enhance accuracy and contribute to computational efficiency in the subsequent stages.

4. Model Development and Training

In this phase, data scientists experiment with various algorithms, architectures, and hyperparameters to create a model that meets the defined objectives. The process includes training models on the prepared data and validating their performance using a separate dataset. Automated Machine Learning (AutoML) tools are increasingly used in this stage to expedite the experimentation process and identify the best-performing models. Version control for both code and model artifacts ensures reproducibility and collaboration among team members.

5. Model Evaluation and Validation

Before a model can move to production, it must be rigorously evaluated to ensure it meets performance requirements. Metrics such as accuracy, precision, recall, and F1 score are commonly used to assess model quality. Beyond these metrics, models are tested for fairness, bias, and robustness under various conditions. Validation techniques, such as cross-validation and holdout testing, are applied to prevent overfitting and ensure the model generalizes well to unseen data.

6. Model Deployment

Once validated, the model is deployed to a production environment to make predictions on live data. Deployment strategies vary depending on the use case and requirements. Options include batch processing for periodic predictions or real-time deployment via REST APIs for instantaneous responses. Continuous integration and continuous deployment (CI/CD) pipelines are key in automating the transition from development to production while minimizing downtime and errors.

7. Monitoring and Maintenance

After deployment, ongoing monitoring ensures that the model continues to perform as expected. Key performance indicators (KPIs) such as prediction accuracy, latency, and system uptime are tracked in real time. Monitoring also involves detecting issues like data drift, concept drift, or anomalies that could compromise the model's effectiveness. Automated alerts and retraining workflows help teams address these issues proactively, maintaining the system's reliability over time.

8. Retraining and Model Iteration

As new data becomes available, models may require retraining to adapt to changing patterns or improve performance. This iterative process involves updating the dataset, retraining the model, and redeploying it with minimal disruption. Predefined thresholds, such as significant changes in data distribution or a drop in model performance, can trigger automated retraining pipelines. These thresholds ensure that the ML system remains up-to-date and aligned with current conditions.

9. Governance and Compliance

Throughout the lifecycle, governance and compliance are critical in ensuring data and models' ethical and lawful use. Ethical and lawful use might include maintaining audit trails, documenting decisions, and adhering to privacy regulations such as GDPR or HIPAA. Explainability and interpretability are also emphasized to build trust with stakeholders and ensure accountability in model predictions.

10. Collaboration and Feedback

The MLOps lifecycle thrives on collaboration between data scientists, engineers, and business stakeholders. Establishing feedback loops allows teams to incorporate user insights and continuously refine the system. Stakeholders can provide valuable input on the model's effectiveness in solving real-world problems, guiding future iterations, and improving.

Core Principles of MLOps

The core principles of MLOps provide a framework for managing the complexities of machine learning operations, ensuring that organizations can effectively build, deploy, and maintain models in production. These principles emphasize collaboration, automation, and adaptability, enabling teams to streamline workflows and achieve business objectives with greater efficiency and reliability. Organizations can create a foundation for sustainable and impactful machine learning practices by adhering to these guiding tenets.

Collaboration is central to MLOps, fostering seamless interaction between data scientists, machine learning engineers, DevOps teams, and business stakeholders. This cross-functional teamwork breaks down silos, ensuring that everyone involved in the machine learning lifecycle has a shared understanding of goals and responsibilities. When teams work together effectively, the insights generated by ML models are more likely to align with organizational objectives and translate into actionable results.

Automation is another cornerstone of MLOps, significantly reducing manual effort in repetitive and error-prone tasks. By automating data preprocessing, model training, and deployment, teams can increase productivity and minimize the risk of mistakes. Automation also extends to monitoring and retraining models, ensuring systems remain accurate and relevant as new data emerges. This focus on automation enables teams to allocate more time to innovation and problem-solving while maintaining operational excellence.

A key aspect of MLOps is the implementation of continuous integration and continuous deployment (CI/CD) practices. These practices ensure that new code, data pipeline updates, and model changes are rigorously tested and deployed efficiently. With CI/CD, organizations can quickly iterate on their ML solutions while maintaining stability and reliability in production environments. This ability to update models seamlessly is crucial for responding to changing business needs and maintaining a competitive edge.

Scalability and flexibility are also integral to MLOps. As machine learning workloads grow in size and complexity, systems must be designed to handle increased demand without compromising performance. Flexibility allows organizations to adapt to new technologies, evolving business priorities, and regulatory changes. Cloud-native platforms, containerization, and orchestration tools like Kubernetes are vital in enabling scalability and adaptability, ensuring that infrastructure and workflows evolve alongside organizational requirements.

Effective monitoring and lifecycle management are essential to maintaining the health and relevance of machine learning models. Monitoring tools track performance metrics, such as accuracy and latency, while detecting issues like data drift or concept drift that can impact model effectiveness. Lifecycle management involves updating or retiring models when necessary, ensuring that production systems remain reliable and aligned with organizational goals. This ongoing oversight is key to sustaining trust in ML systems over time.

Reproducibility is another fundamental principle of MLOps, ensuring that every aspect of the ML lifecycle can be reliably repeated and audited. Version control systems, model registries, and comprehensive documentation enable teams to track changes, reproduce results, and collaborate effectively. Reproducibility fosters transparency, making troubleshooting issues easier, complying with regulations, and building trust in machine learning solutions.

Finally, MLOps emphasizes the importance of delivering tangible business value. Successful machine learning initiatives must address real-world problems and produce measurable outcomes. By focusing on business impact, MLOps ensures that machine learning investments are aligned with strategic goals, driving innovation and long-term success. This principle underscores the importance of connecting technical excellence with organizational priorities to maximize the value of machine learning systems.

Challenges in Implementing MLOps

Implementing MLOps can be a transformative step for organizations seeking to leverage machine learning at scale, but it has its challenges. One of the primary obstacles is the inherent complexity of managing the machine learning lifecycle. Unlike traditional software development, machine learning involves dynamic and evolving elements such as data quality, feature engineering, model tuning, and continuous retraining. Coordinating these interdependent processes requires advanced tools and well-defined workflows, which can be difficult to establish without prior expertise.

Data is often referred to as the foundation of machine learning, but managing data pipelines at scale presents its challenges. Organizations frequently need help with issues related to inconsistent data quality, fragmented sources, and the sheer volume of information. Additionally, ensuring data lineage, governance, and compliance becomes increasingly complicated as data flows through various systems. Addressing these issues demands robust infrastructure and policies, which can be time-consuming and resource-intensive to implement.

Another significant hurdle lies in bridging the gap between data science and operations teams. Data scientists often focus on experimentation and model development, while DevOps teams prioritize scalability, reliability, and deployment. Misaligned priorities and communication breakdowns between these groups can lead to inefficiencies and delays in productionizing machine learning models. Establishing shared goals and fostering cross-functional collaboration are crucial for overcoming this challenge.

Integrating continuous integration and continuous deployment (CI/CD) for machine learning models poses additional difficulties. Traditional CI/CD pipelines are well-suited to static codebases, but machine learning introduces variability through dynamic data and model changes. Creating pipelines that account for data versioning, model evaluation, and retraining without compromising speed or reliability is complex. Teams must invest in automation and testing to ensure seamless integration.

Monitoring machine learning models in production also presents unique challenges. Unlike software systems, where bugs and performance issues can be relatively predictable, machine learning models are susceptible to phenomena like data drift and concept drift. These issues occur when the input data or underlying patterns in the data change over time, leading to degraded model performance. Detecting and addressing these shifts requires sophisticated monitoring tools and a proactive approach to model maintenance.

Reproducibility is another critical aspect that can be difficult to achieve in MLOps. Machine learning workflows often involve numerous steps, from data preprocessing to model training and deployment. Ensuring these processes can be consistently replicated requires meticulous documentation, version control for code and data, and robust tooling. With reproducibility, troubleshooting and auditing have become nearly possible, especially in regulated industries.

Scalability challenges arise as machine learning models are deployed across diverse environments and workloads. Organizations must account for variable demand, ensure infrastructure can handle peak loads, and optimize resource usage to control costs. Implementing scalable solutions often necessitates adopting cloud-native platforms, containerization, and orchestration tools, which can require significant investment and expertise.

Security and compliance represent yet another area of concern. Machine learning systems often process sensitive data, making them attractive cyberattack targets. Additionally, regulatory requirements such as GDPR or HIPAA impose stringent data handling and privacy obligations. Ensuring that MLOps practices comply with these regulations while maintaining robust security protocols is a complex task that requires ongoing attention.

The cost of implementing MLOps can also be a barrier for many organizations. Building and maintaining the necessary infrastructure, hiring skilled personnel, and investing in tools and platforms contribute to substantial upfront and ongoing expenses. Smaller organizations or those new to machine learning may struggle to justify these costs, particularly if the immediate return on investment needs to be clarified.

Another significant challenge is the cultural shift required to adopt MLOps effectively. Teams may be accustomed to working in silos or using ad hoc processes, which can conflict with the collaborative and structured approach of MLOps. Changing mindsets, encouraging teamwork, and promoting a shared understanding of MLOps principles require strong leadership and ongoing effort.

The rapid pace of innovation in machine learning and MLOps tools can also be a double-edged sword. While new technologies promise to simplify workflows and enhance capabilities, staying up-to-date with the latest developments can be overwhelming. Organizations must strike a balance between adopting cutting-edge solutions and maintaining a stable, sustainable MLOps infrastructure.

Finally, defining and measuring success in MLOps can be ambiguous. Unlike traditional software projects, where outcomes are often binary (e.g., the software works or it doesn't), machine learning success involves continuous improvement and nuanced metrics. Organizations must develop clear KPIs for their MLOps initiatives, ensuring they align with broader business objectives. Even technically successful models may only deliver tangible value with this alignment.

By understanding these challenges and addressing them proactively, organizations can unlock the full potential of MLOps, transform their approach to machine learning, and drive meaningful business outcomes.

Versioning in MLOps: Models, Code, and Data

Effective versioning is a cornerstone of MLOps, ensuring that machine learning projects are reproducible, traceable, and adaptable. As machine learning systems grow more complex, managing the relationships between models, code, and data becomes essential for maintaining quality and reliability. Versioning serves as the framework that allows teams to track changes, evaluate performance, and streamline collaboration throughout the machine learning lifecycle.

Versioning models is one of the most critical aspects of MLOps. Models evolve through iterative training and fine-tuning, with each version potentially introducing new features or addressing previous shortcomings. Maintaining a clear record of model versions, including their parameters, architecture, and performance metrics, is vital for auditing and understanding why a specific version was chosen for deployment. With proper model versioning, troubleshooting and improvement efforts can be smooth and efficient.

Code versioning in machine learning projects builds on well-established software development practices. Source control systems like Git enable teams to track changes, manage branches, and collaborate on codebases. However, machine learning introduces additional complexities, such as the need to version scripts for data preprocessing, feature engineering, and model evaluation. Ensuring these scripts are in sync with the corresponding model and data versions is essential for reproducibility and debugging.

Data versioning poses unique challenges not encountered in traditional software development. Data is not static; it evolves as new records are collected, existing entries are updated, or errors are corrected. Capturing and managing these changes through versioning allows teams to trace the lineage of datasets, understand how models were trained, and recreate historical experiments. Technologies and tools like DataOps.live, Delta Lake, or DVC (Data Version Control) are often employed to handle data versioning in MLOps pipelines.

The interplay between models, code, and data underscores the importance of integrated versioning strategies. For instance, a model trained on one dataset version may perform poorly when applied to another, emphasizing the need to tie model versions to the exact data and code used during training. Integrated systems that manage these relationships ensure consistency and prevent mismatches that could lead to errors or degraded performance.

Versioning is also critical for compliance and auditability, particularly in regulated industries. Organizations must often demonstrate how a specific model was developed, including the data on which the model was trained and the decisions behind the model's design. Robust versioning practices make it possible to provide this documentation and address regulatory inquiries confidently. These practices are necessary for compliance efforts to become ad hoc and labor-intensive.

Automating versioning processes is key to scaling MLOps practices. Manual tracking of changes across models, code, and data is time-consuming and prone to human error. Automation tools can log changes, generate metadata, and create snapshots of the entire pipeline, enabling teams to focus on experimentation and optimization rather than administrative tasks.

Collaboration is another area where versioning plays a pivotal role. Machine learning projects often involve cross-functional teams of data scientists, engineers, and business stakeholders. Clear versioning practices help establish a single source of truth, reducing confusion and ensuring that everyone works with consistent and accurate information. As I've described, these practices foster better communication and alignment across the organization.

Additionally, versioning supports rollback and recovery, providing a safety net when changes produce unintended results. Whether it's a model that underperforms in production, a preprocessing script that introduces errors, or a dataset with unexpected anomalies, versioning allows teams to revert to a previous state and quickly diagnose the issue. This capability is invaluable for maintaining system stability and minimizing downtime.

Integrating versioning into CI/CD pipelines further enhances its utility in MLOps workflows. Automated tests can validate changes to models, code, and data, while version control ensures that only approved versions are deployed. This seamless integration reduces the risk of errors and accelerates the deployment process, enabling organizations to iterate and innovate more rapidly.

Despite its benefits, versioning in MLOps comes with challenges. Managing the storage and retrieval of multiple versions, particularly for large datasets and complex models, requires significant infrastructure. Balancing performance and cost while maintaining accessibility can be a delicate task, necessitating careful planning and using scalable storage solutions.

This means that versioning is a foundational practice that underpins the success of MLOps. Organizations can ensure reproducibility, enhance collaboration, and maintain compliance by systematically managing models, code, and data. While the complexities of versioning demand thoughtful strategies and robust tools, its benefits to machine learning operations far outweigh the investment, enabling teams to deliver scalable and impactful machine learning solutions.

Automation in MLOps

Automation is a fundamental pillar of MLOps, enabling teams to streamline machine learning model development, deployment, and management. In an era where machine learning drives innovation across industries, manual workflows are no longer sustainable for scaling operations or maintaining efficiency. Automation reduces human intervention, minimizes errors, and accelerates the iteration cycle, allowing organizations to deploy robust and reliable machine learning systems.

One key area where automation plays a critical role is in the data pipeline. From data ingestion to preprocessing and transformation, automating these steps ensures consistency and repeatability. With tools like Apache Airflow, Kubeflow, and Prefect, teams can schedule and monitor data workflows, ensuring that models are continuously trained on clean and up-to-date data. Keeping clean and up-to-date data also enables rapid retraining of models when new data becomes available, a necessity for applications that rely on real-time insights or frequent updates.

Automated model training and tuning also enhance them. The traditional manual approach of iterating through hyperparameters and testing different algorithms can be time-consuming and error-prone. Automation frameworks, such as Optuna, Hyperopt, or AutoML platforms, allow for efficient hyperparameter optimization and model selection. These tools leverage algorithms to intelligently explore the search space, enabling faster convergence on high-performing models.

Another significant aspect of automation in MLOps is model deployment. Automated deployment pipelines ensure trained models can seamlessly transition from development to production. Continuous Integration and Continuous Deployment (CI/CD) pipelines for machine learning enable teams to test, validate, and deploy models at scale with minimal manual intervention. Tools like Jenkins, GitLab CI/CD, and specialized MLOps platforms like MLflow and TFX enable the automation of these processes while maintaining rigorous quality control.

Monitoring and maintenance are critical for ensuring models perform as expected in production. Automated monitoring systems track key performance indicators (KPIs), such as accuracy, latency, and data drift, alerting teams to potential issues before they impact business outcomes. Tools like Prometheus and Grafana, combined with custom scripts or commercial monitoring solutions, enable the automation of this process. This proactive approach minimizes downtime and helps organizations respond to challenges in real time.

Automation also supports governance and compliance in MLOps workflows. By automatically logging key metrics, data lineage, and model metadata, organizations can ensure auditability and transparency. Automated documentation and tracking systems reduce the compliance burden, particularly in regulated industries such as healthcare, finance, and telecommunications. They also facilitate collaboration by providing a clear model development and deployment process record.

Additionally, automation fosters collaboration between cross-functional teams. Automation allows data scientists, engineers, and business stakeholders to work in parallel by standardizing processes and reducing manual dependencies. Parallelization practices reduce bottlenecks and ensure that resources are used efficiently. For example, automated APIs or shared repositories for data and models make it easier for different teams to access and integrate assets into their workflows.

In summary, automation is indispensable for realizing the full potential of MLOps. It reduces human error, accelerates workflows, and ensures consistency across the machine-learning lifecycle. Organizations can scale their machine learning initiatives by leveraging automation tools and frameworks, responding to dynamic market demands, and delivering reliable and impactful solutions. While the initial investment in automation infrastructure may be significant, the long-term efficiency, scalability, and reliability benefits make it an essential component of any successful MLOps strategy.

Model Deployment Strategies

Model deployment is critical in transitioning a trained machine learning model from a development environment to production, where it can generate real-world value. Selecting the right deployment strategy is pivotal for ensuring performance, scalability, and reliability while addressing the unique requirements of the use case. Various techniques exist, each tailored to specific challenges such as real-time performance demands, resource constraints, and risk tolerance.

One of the most straightforward deployment approaches is direct deployment. Here, the new model replaces the old one in production without intermediate steps. While this method is straightforward and fast, it carries a significant risk: undetected issues in the new model can immediately affect the system. This strategy is typically used in low-risk scenarios or when thorough testing has been completed in a staging environment. Direct deployment is not recommended for critical applications where even brief disruptions could have significant consequences.

For more risk-averse scenarios, blue-green deployment provides a safer alternative. This approach uses two identical environments—blue and green. The current production model runs in the blue environment, while the updated model is deployed to the green environment. Once testing confirms that the new model performs as expected, traffic is routed to the green environment, effectively making it the new production system. This strategy ensures minimal downtime and a quick rollback to the blue environment if issues arise.

Canary deployment takes a more gradual approach to mitigating risk. Instead of fully replacing the old model, a small percentage of user traffic is directed to the new model while the rest continues using the current one. This strategy allows teams to monitor the new model's performance on real-world data before committing to a full rollout. By progressively increasing the traffic to the new model, organizations can validate its reliability while minimizing the impact of potential issues.

Another advanced strategy is shadow deployment. In shadow deployment, the new model runs parallel with the production model but does not affect user-facing operations. Instead, it processes the same input data in the background, and its predictions are compared to the production model's outputs. Shadow deployment is beneficial for testing a model's performance under production-scale loads and identifying potential biases or unexpected behaviors. This approach ensures that the new model is rigorously evaluated without affecting end-users.

For applications requiring multiple models to serve different segments of users or tasks, multi-model deployment can be a viable strategy. This approach involves deploying several models simultaneously, each handling specific aspects of the workload. Load balancers or routing mechanisms direct traffic to the appropriate model based on predefined criteria. While this strategy adds complexity to deployment and monitoring, it can significantly enhance performance and customization for diverse use cases.

Finally, serverless deployment is gaining traction due to its scalability and cost-efficiency. With serverless deployment, models are hosted in cloud environments and triggered by specific events or user requests. Serverless deployment helps to eliminate the need for dedicated infrastructure, allowing organizations to scale up or down dynamically based on demand. Serverless frameworks such as AWS Lambda, Azure Functions, and Google Cloud Functions make deploying models flexibly and cost-effectively easier. However, high-demand scenarios must account for latency considerations and cold-start times.

Choosing the right model deployment strategy requires thoroughly understanding the application's requirements, the organization's risk tolerance, and the resources available. While some strategies prioritize speed and simplicity, others focus on robustness and reliability. By aligning deployment strategies with business goals and operational needs, organizations can maximize the impact of their machine-learning initiatives while minimizing risk and downtime.

Monitoring and Maintenance of ML Models

Monitoring and maintaining machine learning (ML) models in production is essential to ensuring their sustained performance and reliability. Once deployed, ML models interact with dynamic real-world environments where data distributions can shift, user behaviors may change, and external conditions can evolve. With proper monitoring and maintenance, models can quickly improve their accuracy, resulting in better predictions, gain of user trust, and potential business risks. Establishing a robust framework for ongoing oversight helps mitigate these challenges and keeps models aligned with their intended objectives.

One key monitoring aspect is tracking performance metrics such as accuracy, precision, recall, and F1 score. These metrics provide a quantitative measure of how well the model is functioning against both historical and current data. However, it is equally vital to monitor business-specific metrics that indicate whether the model is meeting the organization's broader goals. For instance, consider evaluating a recommendation system based on click-through rates or sales conversions. Reviewing these indicators helps identify when a model's performance starts to drift or degrade.

A significant factor contributing to model degradation is data drift, which occurs when the statistical properties of input data change over time. Data drift can manifest as covariate shift, where the distribution of features changes, or concept drift, where the relationship between inputs and outputs evolves. Detecting these shifts requires implementing automated monitoring tools that compare incoming data distributions to the training data. Alerts can then be triggered when deviations exceed predefined thresholds, prompting further investigation or retraining of the model.

Maintenance activities often include periodic retraining to ensure models stay relevant in changing environments. This process involves incorporating new, up-to-date data into the training pipeline and fine-tuning model parameters. Automated retraining pipelines, a hallmark of MLOps, can significantly reduce the effort and time required for this task. However, retraining must be approached cautiously, as overfitting to recent data or introducing new biases can exacerbate problems rather than solve them. Rigorous validation and testing are necessary to confirm that retrained models perform well across diverse conditions.

Another critical maintenance aspect is ensuring the infrastructure supporting ML models remains robust and scalable. Resource bottlenecks, outdated dependencies, or security vulnerabilities can all disrupt model operations. Regular audits of the production environment and implementing version control and rollback mechanisms help maintain system integrity. Additionally, feedback loops incorporating user interactions and real-world outcomes can provide valuable insights for iterating on model design and improving future performance. By treating monitoring and maintenance as continuous processes rather than afterthoughts, organizations can maximize the longevity and impact of their ML models.

MLOps for Governance and Compliance

MLOps is crucial in ensuring governance and compliance in machine learning operations, especially as regulatory frameworks around AI and data become more stringent. With the increasing use of ML models in sensitive areas like healthcare, finance, and criminal justice, organizations must adhere to regulatory requirements to avoid legal and financial consequences. MLOps provides the structure and processes necessary to integrate compliance measures into the lifecycle of ML models, ensuring transparency, accountability, and ethical use of data and algorithms.

One of the core elements of MLOps for governance is traceability. In a regulated environment, it's essential to maintain a detailed record of all data, model versions, and the decisions made throughout the ML pipeline. Regulated environment artifacts might include documenting how data was collected, processed, and used for training and changes made to models during development and deployment. Automated tracking systems can log this information in real-time, creating an immutable audit trail. These audit trails allow organizations to demonstrate compliance during audits and ensure that models can be rolled back to a previous state if issues arise, mitigating risks associated with model errors or biased predictions.

Data privacy and security are also paramount in governance. MLOps frameworks incorporate secure data handling practices, such as anonymization and encryption, to protect sensitive information in training and inference. In some industries, like healthcare (e.g., HIPAA in the United States) or finance (e.g., GDPR in the EU), strict laws govern how data should be protected. MLOps helps enforce these regulations by integrating privacy-by-design principles into the ML pipeline, ensuring that personal data is only accessible to authorized individuals and that models do not inadvertently leak confidential information. Regular checks can also be automated to verify that models comply with these privacy regulations throughout their lifecycle.

In addition to privacy and traceability, fairness is another critical aspect of governance in MLOps. Models must be continuously monitored for biases that could result in unfair or discriminatory outcomes. MLOps frameworks provide tools to assess model performance across different demographics, ensuring that predictions are equitable and do not reinforce societal disparities. Examples of societal disparities may include implementing bias detection during training and deployment phases and adopting strategies such as fairness constraints and adversarial testing. Governance structures built within MLOps platforms help organizations take proactive steps to address these ethical concerns, minimizing the risk of regulatory violations or reputational damage.

Lastly, model explainability is an essential component of MLOps for governance and compliance. Many industries, particularly those with high-stakes decisions, require ML models to be interpretable and transparent. MLOps frameworks facilitate the creation of interpretable models and tools that provide insights into how models make predictions. This facilitation helps build trust with stakeholders and allows organizations to justify their automated decisions to regulators. By embedding explainability into the model development and deployment process, MLOps ensures that organizations can maintain compliance with regulations requiring transparency in AI decision-making.

Case Studies and Real-World Applications

MLOps has rapidly become a cornerstone for organizations looking to scale their machine learning initiatives efficiently and responsibly. Real-world applications and case studies highlight the benefits of adopting MLOps practices in various industries. One of the most notable examples is in the finance sector, where MLOps has streamlined model deployment and improved decision-making. For example, banks have implemented MLOps pipelines to automate fraud detection models, enabling them to monitor transactions and detect suspicious activity continuously in real time. By using MLOps, financial institutions ensure that these models are updated regularly, tested for performance, and comply with regulatory requirements, all while minimizing the manual intervention needed.

In healthcare, MLOps is playing a critical role in transforming patient care by enabling the deployment of machine learning models for diagnostic tools and predictive analytics. Hospitals and healthcare organizations increasingly rely on predictive models to assist in diagnosing diseases early, predicting patient outcomes, and personalizing treatment plans. By adopting MLOps, healthcare providers can efficiently manage the lifecycle of these models, ensuring that they are consistently trained on the latest data and perform accurately across different patient populations. Furthermore, MLOps helps mitigate risks associated with data privacy, as automated workflows enforce compliance with regulations like HIPAA in the United States and GDPR in Europe, ensuring that sensitive patient data is properly handled.

The retail industry has also seen significant advantages from MLOps, particularly in personalized marketing and demand forecasting. Retailers use machine learning to analyze vast amounts of customer data to recommend customized products and optimize inventory management. By implementing MLOps pipelines, these companies can continuously deploy updated models that respond to changing customer preferences and market trends. One case study involves a global e-commerce platform that uses MLOps to maintain recommendation systems that adapt in real time to customer behavior, improving user experience and sales conversion rates. Automating model deployment and testing ensures that the recommendations remain relevant, and using version control ensures that the models are both reproducible and accountable.

In the manufacturing sector, MLOps is being used to optimize production processes and predictive maintenance. Manufacturers are using machine learning to analyze sensor data from equipment to predict failures before they occur, allowing them to minimize downtime and extend the lifespan of machinery. One notable case is a car manufacturer that adopted MLOps to monitor the performance of its assembly line robots. By automatically deploying and updating predictive models that detect anomalies in robot behavior, the company reduced unexpected maintenance costs by 30%. MLOps practices ensured that the predictive models were integrated seamlessly into the operational workflow, continuously tested, and monitored for accuracy, enabling the manufacturer to scale its efforts across multiple plants.

Telecommunications companies have also embraced MLOps for network optimization and customer service improvements. For instance, a leading telecom provider uses machine learning to predict network congestion and dynamically adjust resource allocation. By adopting MLOps, the company can maintain model quality across a large-scale network, ensuring that new models are deployed swiftly without interrupting services. MLOps also enables the telecom provider to track and monitor the performance of customer service chatbots, adjusting their models to understand better and respond to customer inquiries. This combination of predictive analytics and continuous model maintenance has led to higher customer satisfaction and reduced churn.

The energy sector, particularly in the context of renewable energy, is another area where MLOps is being leveraged. One such example is an energy company using machine learning to predict energy demand and optimize the distribution of renewable energy from solar and wind sources. By integrating MLOps into its infrastructure, the company ensures that its models are regularly updated with new data and can predict energy consumption patterns accurately. MLOps allows for rapid deployment of models that adjust to fluctuations in weather patterns, making renewable energy distribution more reliable and efficient. Furthermore, MLOps practices help comply with environmental regulations, ensuring that the models used in the energy grid adhere to sustainability goals and regulatory standards.

Finally, the automotive industry has adopted MLOps for autonomous vehicle development and improvement. Companies like Tesla and Waymo have built MLOps pipelines to continuously test, train, and deploy models that power self-driving cars. These models must be updated frequently with data from real-world driving experiences to improve their performance in diverse environments. Through MLOps, automotive companies can manage vast amounts of sensor data, validate models, and deploy improvements rapidly. These data points have accelerated the development of autonomous driving technologies and have enabled these companies to respond quickly to safety concerns or regulatory requirements, ensuring that their models are safe, effective, and compliant with industry standards.

These case studies illustrate how MLOps transforms industries by streamlining model deployment, improving scalability, and ensuring regulatory compliance. Adopting MLOps across sectors such as finance, healthcare, retail, manufacturing, telecommunications, energy, and automotive highlights its versatility and wide-reaching benefits. Organizations can stay agile and responsive by automating the processes around model development, deployment, and monitoring, reducing operational costs, improving decision-making, and driving innovation.

Conclusion

MLOps represents a critical evolution in how organizations develop, deploy, and maintain machine learning models. As machine learning becomes an increasingly integral part of modern business operations, the need for streamlined, automated processes to manage the lifecycle of these models has never been more apparent. MLOps facilitates faster deployment and continuous improvement of models and ensures that they are scalable, reliable, and compliant with industry standards and regulations. By embracing MLOps, organizations can better integrate machine learning into their workflows, optimizing decision-making, enhancing operational efficiency, and driving innovation.

Ultimately, MLOps provides a framework that bridges the gap between data science, IT operations, and business goals, fostering collaboration and ensuring that models deliver consistent value. The case studies and real-world applications discussed throughout this chapter underscore the immense potential of MLOps in various sectors, from healthcare and finance to manufacturing and telecommunications. By addressing the challenges of versioning, automation, model deployment, and monitoring, MLOps enables organizations to scale their machine learning initiatives and maintain a competitive edge in an increasingly data-driven world.

CHAPTER 4

DataOps Best Practices

DataOps, much like its counterpart DevOps in software engineering, emphasizes collaboration, automation, and integration to streamline the data management lifecycle. However, it focuses on the challenges and complexities inherent in the data landscape. In this chapter, we will explore best practices that enable organizations to build, deploy, and maintain robust data pipelines, ensuring that data is accurate, accessible, and usable across different teams. By incorporating the principles of DataOps, organizations can overcome the traditional bottlenecks and inefficiencies associated with data operations, leading to faster, more reliable data delivery and enhanced decision-making.

As DataOps continues to evolve, establishing a set of best practices is key to ensuring successful implementation and sustainable growth. Whether managing data pipelines, ensuring data quality, or fostering collaboration between cross-functional teams, understanding and applying these best practices can significantly enhance the value of data within an organization. In this chapter, we'll explore practical approaches to creating a well-governed, scalable, and adaptable data ecosystem that meets businesses' growing needs.

Establishing Cross-Functional Teams for Data Collaboration

Establishing cross-functional teams is one of the cornerstones of DataOps, as it enables seamless collaboration between diverse departments within an organization. Traditionally, data operations have been siloed, with data engineers, analysts, scientists, and other stakeholders working in isolation. This fragmentation can lead to communication gaps, slow decision-making, and inconsistent data outputs. By creating cross-functional teams, organizations can foster better communication, streamline workflows, and ensure that each team member's expertise is integrated into the overall data pipeline. These teams work collaboratively to design, build, and maintain data systems, driving a more agile and adaptive approach to data management.

R. L. Steelman, Jr., *Mastering Snowflake DataOps with DataOps.live*, https://doi.org/10.1007/979-8-8688-1754-0_4

A key component of these cross-functional teams is ensuring each member understands their role and the broader data ecosystem. Data engineers might focus on infrastructure and data pipelines, while data analysts and scientists can bring insights into how data should be transformed, analyzed, and visualized. Furthermore, stakeholders from business teams should be part of the discussions, ensuring that data operations align with strategic goals and deliver meaningful business value. When all team members align on a shared vision, the result is more efficient and effective data management, leading to better outcomes and quicker turnaround times.

In a cross-functional team setup, communication and collaboration are paramount. Regular meetings, transparent processes, and shared platforms for tracking progress are essential to ensure all teams are on the same page. The integration of tools for project management, version control, and data cataloging can help break down silos and improve the flow of information. Collaborative platforms allow teams to see the status of data pipelines, track ongoing analyses, and quickly address issues as they arise. Collaboration at this level also fosters a culture of continuous improvement, with established feedback loops, and team members can learn from each other's experiences and expertise.

Another key advantage of cross-functional teams is the ability to address data quality issues proactively. In traditional data management models, quality control is often an afterthought, only addressing problems when they arise. In a cross-functional setup, however, data quality becomes a shared responsibility. For example, data engineers and data scientists can work together to define and enforce data validation rules, ensuring that data is accurate and reliable as it moves through the pipeline. This collaborative approach helps catch issues early and prevents them from snowballing into larger, more costly problems.

Moreover, cross-functional teams enable more agile data operations, essential in today's fast-paced business environment. With teams working in tandem, changes to data requirements, new data sources, or adjustments to existing pipelines can be implemented more swiftly. This agility also supports the iterative nature of DataOps, where pipelines are continuously refined, tested, and optimized. Teams can quickly adapt to new challenges, adjust to evolving business needs, and implement new technologies or methodologies without disrupting the entire system.

Finally, leadership plays a critical role in ensuring the success of cross-functional teams. Leaders must encourage a culture of collaboration, set clear expectations, and provide the necessary resources and tools to enable teams to work effectively. A culture like the one I describe might include fostering a safe environment for experimentation, where team members feel empowered to take risks and innovate without fear of failure. By supporting cross-functional teams with the right resources and leadership, organizations can maximize the benefits of DataOps and unlock the full potential of their data-driven initiatives.

Collaboration Tools for Enhancing Cross-Functional Teamwork

To fully realize the benefits of cross-functional teams in DataOps, organizations must leverage collaboration tools that facilitate seamless communication and workflow integration. Even the most well-structured teams risk inefficiencies, misaligned goals, and lost productivity without adequate tools. These tools serve as a central hub where engineers, analysts, scientists, and business stakeholders can interact, ensuring that data workflows remain transparent and accessible to all relevant parties. By implementing the right collaboration platforms, organizations can reinforce the principles of DataOps and drive continuous improvement.

One critical type of collaboration tool is project and workflow management platforms. Solutions such as Jira, Trello, or Asana enable teams to track progress on data pipeline development, model deployment, and issue resolution. These platforms provide visibility into ongoing work, assign responsibilities, and set deadlines, reducing miscommunication and improving accountability. Automated notifications ensure that real-time updates are shared, keeping all team members aligned and responsive to changes.

In addition to workflow tracking, version control and data cataloging tools are crucial in maintaining consistency and integrity within cross-functional teams. Platforms like Git, DVC (Data Version Control), and Snowflake's Data Marketplace enable teams to manage data assets efficiently while ensuring reproducibility and governance. By maintaining a single source of truth, these tools help prevent discrepancies in data interpretation and usage. This structured approach to data versioning and accessibility fosters greater trust in data-driven decision-making.

Real-time communication and documentation platforms further enhance collaboration by streamlining discussions and knowledge sharing. Tools like Slack, Microsoft Teams, and Confluence provide a structured space for problem-solving, brainstorming, and maintaining a shared repository of insights and best practices.

These platforms help bridge gaps between technical and business teams, ensuring that strategic objectives remain at the forefront of DataOps initiatives. When used effectively, collaboration tools improve efficiency and reinforce a culture of shared responsibility and innovation within data-driven organizations.

Automating Data Pipelines for Consistency and Speed

Automating data pipelines is crucial for achieving consistency and speed in modern data operations. In a manual data pipeline, data extraction, transformation, and loading (ETL) are often time-consuming and prone to human error. With the increasing volume, variety, and velocity of data, manual processes become insufficient for handling large-scale, complex data environments. Automation allows you to perform these processes reliably and at scale, ensuring that data is processed consistently without the delays associated with human intervention. Automation enables faster data delivery to stakeholders and accelerates decision-making by removing the bottlenecks introduced by manual steps.

One of the primary advantages of automating data pipelines is the reduction of errors that can occur in manual processes. Manual interventions often lead to inconsistencies, such as incorrect data formatting, failed transformations, or missed data integration steps. These errors can propagate through the pipeline, causing complex downstream issues to diagnose and fix. Automating the data flow mitigates these risks, and data processing becomes more reliable. Automated tests can be introduced at various pipeline stages, ensuring that each data transformation or integration step performs as expected and catches errors before they impact the business.

In addition to ensuring consistency, automation significantly improves the speed at which data flows through the pipeline. Automated workflows eliminate the need for manual approvals and interventions, allowing data to move from raw sources to actionable insights faster. This speed is essential in environments that require real-time or near-real-time analytics, such as fraud detection systems or customer sentiment analysis. Automated pipelines can process vast amounts of data rapidly, allowing organizations to react to changes in data as they happen. This enhanced speed also enables teams to iterate on data models rapidly, experiment with new data sources, and refine existing processes without slowing down the overall pipeline.

To achieve true automation, it is essential to implement robust orchestration tools. These tools manage the execution and coordination of various tasks within the pipeline, ensuring that processes are executed in the correct order and that dependencies are appropriately handled. Orchestration platforms, such as Apache Airflow or Prefect, provide the infrastructure necessary to schedule, monitor, and log data pipeline executions. They also facilitate error handling, allowing teams to set up retries or notifications when things go wrong, reducing the need for manual oversight. By using these tools, teams can create a smooth and predictable flow of data that adapts to the complexity of the task.

Beyond improving speed and consistency, automating data pipelines also significantly boosts scalability. As businesses grow, the volume of data they need to process increases, and manual systems simply need to catch up. Automated pipelines, however, can easily scale to accommodate additional data sources, greater data volume, or more complex transformations. Automation allows teams to handle growing data demands without requiring proportional resource increases or effort. This scalability ensures that their data infrastructure can grow as organizations expand, supporting larger datasets, more intricate analytics, and a broader array of use cases.

Automation also enhances transparency and governance within data pipelines. Every automated step can be logged, making it easier to track how data flows through the system and where issues arise. Raising these issues creates an auditable trail essential for compliance with data governance policies. Automated pipelines also make it easier to implement version control and rollbacks, ensuring that teams can revert to previous versions of the pipeline or data transformations when necessary. With these capabilities, teams can maintain complete control over data processes while ensuring that the best data governance and security practices are always followed.

Data Quality and Testing: Ensuring Reliability at Scale

Data quality is a cornerstone of any successful data operation, and ensuring its integrity is particularly critical as organizations scale their data processing capabilities. As data flows through various stages in a pipeline, from collection to transformation and storage, maintaining high data quality becomes increasingly complex. With large datasets and frequent updates, minor errors can have significant consequences, from misleading analytics to incorrect business decisions. Thus, data quality management must be integral to the data pipeline, ensuring that data is accurate, consistent, and reliable at all stages.

Testing is one of the most effective ways to ensure data quality. We design automated data tests to catch errors before having the opportunity to propagate through the pipeline, helping prevent data issues from reaching the business. These tests can check for everyday situations like null values, incorrect data types, outliers, or duplicate entries. They can also verify that data transformations, such as aggregations or joins, are performed correctly. By embedding these tests into the pipeline, teams can monitor the health of their data in real time and quickly address any discrepancies before they impact the decision-making process.

One of the key challenges in ensuring data quality at scale is dealing with the complexity of modern data environments. Data may come from various sources, including internal systems, third-party providers, and user-generated content, resulting in inconsistencies or mismatches in formats, values, or structures. Automated testing can address these challenges by providing a standardized approach to checking data quality, regardless of source. As data pipelines grow and evolve, it is important to continuously review and refine the tests, ensuring they cover new sources, transformations, and business logic.

Data validation is another important aspect of data quality assurance. Validating data ensures that the data conforms to specific business rules or expectations, preventing anomalies in data affecting downstream analysis. For example, validating that a customer's age is within a reasonable range or that an order's total amount matches the sum of its line items ensures that only trustworthy data enters the pipeline. Building these validations into automated processes ensures that they are applied consistently and without delay, reducing the risk of errors going unnoticed.

As organizations scale their data pipelines, manual testing becomes increasingly impractical. The sheer volume of data and complexity of the transformations make it impossible for human testers to monitor every aspect of the system. Thus, automation plays a crucial role. By automating data testing at various stages of the pipeline, teams can maintain a high level of assurance about the quality of their data without requiring additional manual effort. Automated testing frameworks can run tests at regular intervals, offering real-time feedback on the health of the data and even alerting teams to issues as they arise.

Data quality testing also needs to account for the evolving nature of data. Data schemas and structures can change, especially in agile environments where data products are continuously being developed or modified. To keep up with these changes, the testing framework should be dynamic and be able to adapt to new data formats, models, or sources. For example, adding a new data source may require updates to the tests that validate data from that source, ensuring that they align with the new schema. By maintaining flexible and up-to-date testing practices, organizations can ensure their data pipelines remain robust even as their data landscape evolves.

Another critical consideration for data quality at scale is handling large volumes of data in real time. Data must be processed and analyzed instantly in many use cases, such as financial fraud detection or real-time recommendation systems. Such critical systems require rigorous testing to ensure accurate and relevant real-time data processing. Automated tests can help monitor data flow at high velocity, ensuring that anomalies are quickly flagged and addressed without disrupting the overall flow of operations. Real-time data quality monitoring ensures that critical systems remain operational and reliable under the pressure of large, fast-moving datasets.

Lastly, data quality testing should be part of a broader data governance strategy. As organizations grow and handle more sensitive or regulated data, they must demonstrate compliance with privacy and security standards. Data quality testing supports this by ensuring that sensitive data is protected, valid, and used appropriately. For example, automated tests can verify that data retention policies are adhered to, ensuring the retirement of obsolete or unnecessary data. By embedding testing within a comprehensive governance framework, organizations can improve data quality and ensure compliance with legal and regulatory requirements. This approach builds a solid foundation for reliable, secure, and compliant data operations.

Managing Data Governance and Compliance in DataOps

Managing data governance and compliance in a DataOps environment is essential for ensuring data is handled securely and responsibly throughout its lifecycle. Organizations must ensure that their data operations adhere to these legal frameworks in a world of increasing data regulations, such as GDPR, CCPA, and HIPAA. Data governance establishes the rules and processes for access, ownership, privacy, and security. It ensures that the right people can access the right data at the right time while maintaining its integrity and confidentiality. A well-defined governance structure within DataOps can help organizations ensure compliance with both internal policies and external regulatory requirements.

One of the primary challenges in DataOps governance is managing data privacy across different teams and environments. With data often being shared across multiple departments, stakeholders, and even third-party vendors, tracking how users access, modify, and store data becomes crucial. Such detailed monitoring levels require strict data access controls and audit trails that can track data movement throughout the pipeline. Automated tools can assist in this process by monitoring who is accessing what data and flagging any unauthorized attempts. Data masking, encryption, and anonymization techniques can also be applied to sensitive data to ensure privacy while enabling collaboration across teams.

Another important aspect of managing governance in DataOps is ensuring you maintain data quality and compliance as pipelines scale. As data sources proliferate and the volume of data grows, the complexity of ensuring data accuracy, consistency, and lineage increases. With automated testing integrated into the pipeline, teams can continuously verify that data conforms to governance standards and meets compliance requirements. These tests include checks for missing data, incorrect data types, or business rule violations, which could lead to regulatory or security risks if left unaddressed. By embedding governance checks into the pipeline itself, organizations can proactively address issues before they escalate.

Incorporating compliance monitoring into the DataOps workflow helps organizations stay agile while maintaining control over their data. Rather than relying on periodic, manual audits, automated compliance checks can continuously monitor the flow of data and ensure that it adheres to regulatory standards. These checks allow organizations to respond to compliance challenges in real time, adjusting their data practices as necessary to avoid penalties or reputational damage. DataOps must remain flexible as data regulations evolve, ensuring that the necessary controls and audit mechanisms are always in place to support changing compliance requirements. By aligning governance with DataOps principles, organizations can ensure that their data operations remain secure, compliant, and efficient.

Version Control and Data Lineage: Tracking Changes Effectively

Version control and data lineage are critical elements in ensuring data integrity, traceability, and quality within a DataOps environment. Version control, often associated with software development, involves tracking changes to code, configurations, and data throughout its lifecycle. In the context of DataOps, you should apply version control to the codebase and the data itself. This approach allows teams to manage and document changes to datasets, data transformations, and model versions. By implementing version control systems for data, organizations can quickly revert to previous versions, track the evolution of datasets, and maintain consistency across various environments.

On the other hand, data lineage is the visual or documented representation of the flow of data through systems, processes, and transformations. It provides a comprehensive view of how data moves, is transformed, and is used throughout an organization's systems. Understanding data lineage is crucial for ensuring data quality and compliance, enabling teams to trace data back to its source and verify its authenticity. It also helps identify pipeline bottlenecks, errors, or inefficiencies, allowing teams to take corrective action before issues affect downstream processes. Combining version control with data lineage offers a complete view of the data and its transformation history.

When managing data lineage and version control, it's essential to establish transparent processes for tracking changes. Every modification to data, whether it's an update to the data source, a change in transformation logic, or an alteration to how data is being accessed, should be documented and tracked. This documentation provides context for each change, allowing teams to understand the impact of those changes on downstream processes and ensuring that the data remains accurate and consistent. Moreover, these records serve as a vital resource for troubleshooting and understanding the root causes of data issues. As organizations scale their data operations, it becomes increasingly difficult to keep track of these changes manually, which is why automating data lineage and version control is vital for maintaining efficiency and reducing human error.

Automating version control and data lineage helps to ensure that data is consistently tracked and auditable. Tools that integrate with the DataOps pipeline can automatically capture every change, from data ingestion to transformation, and document it for future reference. These automated systems can also maintain an immutable record of changes, which can be essential for regulatory compliance. For instance, financial institutions or healthcare organizations must maintain strict data audit trails. Automated version control and data lineage solutions efficiently manage and verify these audit requirements while reducing the manual workload associated with compliance.

The combination of version control and data lineage also supports collaborative efforts among teams. Data engineers, analysts, and scientists can work more efficiently by accessing accurate records of the data's history and transformation journey.

This transparency enables teams to understand how data is structured, how it has evolved, and how changes in one part of the pipeline might affect other parts of the system. Collaboration is further enhanced when changes are tracked in real-time, allowing teams to stay aligned and quickly resolve any issues or discrepancies. Additionally, it ensures that everyone is working with the correct version of the data, minimizing the risk of using outdated or incorrect data in analyses or models.

Finally, version control and data lineage are essential for building trust in the data. For stakeholders, knowing that they can trace the data to its source and view its transformation history assures that the data is reliable and accurate. This transparency fosters confidence in the insights derived from the data and allows decision-makers to make informed choices. Without a clear view of the data's journey, it's difficult to assess its quality and reliability, which can undermine the credibility of any analysis or model. By incorporating version control and data lineage into the DataOps pipeline, organizations improve their operational efficiency and reinforce the trustworthiness of their data.

Scaling DataOps with Cloud and Hybrid Architectures

Scaling DataOps with cloud and hybrid architectures offers organizations the necessary flexibility and scalability to handle growing data volumes and increasing complexity in their data environments. Cloud architectures provide an on-demand infrastructure, enabling organizations to quickly scale their data pipelines without requiring large upfront investments in physical hardware. With cloud platforms like AWS, Azure, and Google Cloud, DataOps teams can leverage advanced tools and services, like data storage, processing power, and machine learning capabilities, which can be easily scaled as demands increase. The cloud's ability to scale resources dynamically means teams can focus on delivering data solutions without worrying about infrastructure limitations.

Hybrid architectures, which combine both on-premise and cloud solutions, offer additional flexibility for organizations with specific compliance or data residency requirements. Data security and privacy are critical concerns in many industries, and specific sensitive data must be kept within the organization's infrastructure. Hybrid environments allow teams to process and store sensitive data on-premises while utilizing the cloud for less sensitive data or additional computational power. This approach ensures that organizations can utilize cloud scalability while maintaining control over specific aspects of their data operations, optimizing security and performance.

One of the key advantages of scaling DataOps in the cloud is the ability to integrate various data services and platforms that enhance pipeline automation. With cloud-native tools, teams can automate real-time data ingestion, processing, and monitoring, allowing for more efficient workflows and quicker decision-making. Cloud providers offer services targeting the optimization of data pipelines, including data lakes, managed ETL (Extract, Transform, Load) tools, and data warehousing platforms. These services enable DataOps teams to process large datasets in parallel, ensuring faster data flow and reducing bottlenecks that typically arise from on-premise limitations.

For organizations using hybrid architectures, the challenge lies in maintaining consistency across both on-premise and cloud environments. Hybrid environments require robust data orchestration tools that ensure seamless integration between the two. These tools include synchronizing data between on-premise databases and cloud storage, ensuring that all data is accurately mirrored and updated. Additionally, hybrid architectures demand careful management of governance and security policies to ensure that data remains compliant with internal and external regulations regardless of where it is stored. While hybrid setups introduce additional complexity, they also enable organizations to optimize cost, security, and performance to suit their unique needs.

Finally, as organizations scale their DataOps practices, cloud and hybrid architectures enable better collaboration and agility. Cloud platforms provide centralized data access, allowing cross-functional teams, including data engineers, analysts, and data scientists, to collaborate more effectively, regardless of physical location. The cloud's flexibility also supports continuous integration and continuous delivery (CI/CD) processes, ensuring that teams can rapidly test, deploy, and iterate on data products and analytics. This improved collaboration accelerates the development of new data-driven insights and innovations, ultimately helping organizations stay competitive in an increasingly data-driven world.

Conclusion

As organizations continue to scale their data operations, leveraging cloud and hybrid architectures becomes critical in achieving agility and efficiency. These architectures provide the flexibility to accommodate growing data volumes while ensuring compliance, security, and performance. By integrating advanced cloud-native tools, teams can automate data pipelines and enhance collaboration, resulting in faster, more accurate data insights. Hybrid environments further allow organizations to maintain control over sensitive data, balancing security needs with the scalability benefits of the cloud. This balance enables businesses to build robust, cost-effective DataOps ecosystems that evolve with their changing needs.

Your ability to scale DataOps practices through cloud and hybrid solutions positions your organization to meet the demands of the modern data landscape. By implementing these architectures, businesses can streamline data workflows, ensure high-quality governance, and improve cross-team collaboration. Combining scalability, automation, and integration across environments helps teams tackle data challenges at scale, leading to more informed decision-making and a competitive edge in today's data-driven market. Ultimately, embracing cloud and hybrid architectures equips organizations with the tools to innovate and adapt to the evolving data ecosystem.

CHAPTER 5

Understanding Snowflake

In this chapter, we will explore Snowflake, a cloud-based data platform that has revolutionized how organizations approach data warehousing, analytics, and data sharing. Known for its ability to scale efficiently, support high-performance queries, and provide flexible data storage, Snowflake has become a key player in the data ecosystem. This chapter will comprehensively review Snowflake's architecture, key features, and how it stands out in data warehousing.

By delving into Snowflake's core components, users will better understand how its unique architecture facilitates the seamless integration of structured and semi-structured data. We will cover its powerful features, such as automatic scaling and data sharing, and look into the benefits organizations can expect when implementing Snowflake in their data environments. This exploration will lay the foundation for understanding how Snowflake is designed to meet the demands of modern data workflows and enhance overall business performance.

Overview of Snowflake Architecture and Features

Snowflake is a cloud-native data platform leveraging design principles to handle various data processing tasks, including data warehousing, data lakes, and analytics. The architecture was built to take full advantage of cloud scalability, offering a flexible and efficient system that automatically adjusts to different workloads. Snowflake operates on top of cloud infrastructure services like AWS, Microsoft Azure, and Google Cloud, providing a seamless integration with the broader cloud ecosystem. This cloud-based foundation enables users to eliminate the complexities of managing on-premises hardware, offering both elasticity and cost-effectiveness that are unmatched by traditional data systems.

R. L. Steelman, Jr., *Mastering Snowflake DataOps with DataOps.live*, https://doi.org/10.1007/979-8-8688-1754-0_5

One of the most significant features of Snowflake's architecture is its separation of compute and storage layers. Unlike traditional data warehouses, where storage and compute are tightly coupled, Snowflake allows users to scale storage and compute independently. This level of scaling control enables businesses to allocate computing resources based on their current workload requirements, reducing costs during low-demand periods and providing the necessary power during heavy workloads. Snowflake shares the storage layer among all users and automatically scales to accommodate both structured and semi-structured data, including JSON and Parquet.

The multi-cluster shared data architecture of Snowflake is another standout feature. Snowflake's architecture allows multiple compute clusters to access the same data simultaneously without impacting performance. This architecture ensures that different teams, departments, or workloads can run in parallel without interfering with each other, making it ideal for large organizations with multiple users. The system uses a central repository to store all data, making it easy to share and collaborate without duplicating data. Additionally, Snowflake's unique approach allows automatic scaling of compute resources to match workload demands, ensuring optimal performance at all times.

The design principles for Snowflake's elastic compute layer allow it to automatically scale up and down based on workload needs, further enhancing its performance and efficiency. This elasticity is especially beneficial in handling fluctuating or unpredictable workloads. Snowflake can quickly provide additional virtual computing resources to handle the demand when a query or data transformation is required. Once the task is complete, Snowflake automatically deallocates resources, meaning companies only pay for the compute they use. This capability makes Snowflake highly cost-effective, as businesses can avoid over-provisioning and only pay for what they need, helping control real-time costs.

Snowflake's native support for semi-structured data is also one of its key features. Traditional data warehouses primarily store structured data, which requires modeling the data into predefined schemas. In contrast, Snowflake's design principles allow structured and semi-structured data, making it suitable for modern data ecosystems that deal with large amounts of diverse data types. Snowflake automatically parses and stores semi-structured data in its VARIANT format, making it easier to load, query, and analyze complex data such as JSON, XML, or Parquet without requiring complex data transformation.

Another defining feature of Snowflake is its built-in security and governance capabilities. Snowflake offers enterprise-grade encryption at rest and in transit, ensuring data is securely stored and transmitted. It also supports role-based access controls (RBAC), allowing administrators to define granular user permissions to ensure only authorized users can access sensitive data. Snowflake's automated data protection features help organizations maintain compliance with various regulatory standards, including GDPR, HIPAA, and SOC 2, making it a reliable choice for industries that require strict data security and governance.

These architectural features make Snowflake a versatile, efficient, and secure platform for modern data management. Its ability to scale resources, handle diverse data types, and ensure high performance across multiple users and workloads enables organizations to simplify their data operations while reducing complexity and costs. Snowflake's cloud-native approach ensures that businesses can focus on leveraging their data for strategic decision-making without worrying about the underlying infrastructure.

Key Concepts of Snowflake

Snowflake's architecture revolves around several key concepts differentiating it from traditional data warehousing solutions. At its core, Snowflake uses a multi-cluster shared data architecture, which separates compute and storage layers. This separation allows for independent scaling of resources, enabling businesses to optimize costs and performance. With this approach, users can provision compute power based on workload demands while the storage layer remains unified and shared across all users. This flexible architecture ensures that performance is not compromised, even when multiple users or teams work simultaneously with large datasets.

Another key concept in Snowflake is virtual warehouses, serverless clusters of compute resources that Snowflake automatically scales. Virtual warehouses are independent and can be sized according to the workload's needs, offering elastic scaling. Users can run multiple virtual warehouses on the same data set without performance degradation, ensuring that each team or user gets dedicated resources without interfering with others. The ability to pause and resume these warehouses further optimizes costs, as businesses are only billed for active compute time.

Snowflake organizes its underlying structure into databases, schemas, and tables, all similar to traditional data warehousing systems. However, Snowflake also supports semi-structured data such as JSON, XML, and Parquet through its VARIANT data type.

VARIANT data types make it a versatile solution for handling both structured and unstructured data in a unified manner. Data in Snowflake is automatically parsed and queryable using SQL without the need for complex transformations or custom parsing processes. Parsing data in this manner eliminates the need for preprocessing and provides flexibility in data handling.

Data sharing is another fundamental concept in Snowflake. It enables organizations to securely share data with other organizations or departments without creating duplicate copies. Data sharing is particularly important in companies with multiple stakeholders, as data can be available for real-time analysis while maintaining centralized control and governance. Snowflake's data-sharing capability operates seamlessly, allowing recipients to access the data as if they were storing it locally without needing to manage the underlying infrastructure.

Security and governance underscore the entirety of Snowflake's architecture, with features like role-based access control (RBAC) and end-to-end encryption. Snowflake uses RBAC to manage user permissions and ensure that only authorized individuals can access sensitive data. Robust RBAC enhances security while providing fine-grained control over data access. Additionally, Snowflake supports automated backups, failover, and data replication, ensuring that data is always available and protected. These features, combined with its robust monitoring and auditing capabilities, make Snowflake a reliable choice for organizations looking to maintain strong data governance and comply with regulatory standards.

These key concepts form the foundation of Snowflake's powerful and flexible platform, making it a suitable solution for modern data management. Its ability to separate compute and storage, handle both structured and semi-structured data, and provide seamless data sharing and security features sets it apart from traditional data warehouses. It makes it an attractive choice for organizations exploring options that enable them to leverage their data at scale.

Benefits of Snowflake for Data Warehousing

Snowflake offers several compelling benefits, making it a strong choice for data warehousing. One of the most significant advantages is its elastic scalability, which allows businesses to scale up or down based on demand easily. This flexibility ensures that organizations only pay for the compute resources based on use, eliminating the need for overprovisioning. Whether running large, resource-intensive queries or smaller, more frequent operations, Snowflake dynamically adjusts to optimize costs and performance, making it ideal for businesses with fluctuating workloads or variable data processing needs.

Another key benefit of Snowflake is its ability to handle both structured and semi-structured data in a unified platform. Traditional data warehouses often require separate systems or tools to manage semi-structured data like JSON, XML, and Parquet. Snowflake, however, integrates these data types seamlessly alongside structured data, enabling users to query and analyze them together. This reduces complexity in data management and allows organizations to work with all types of data without needing additional infrastructure or custom solutions.

Snowflake also automatically scales compute resources, meaning that workloads get dynamically allocated to different virtual warehouses based on demand. Dynamic allocation of virtual warehouses eliminates the manual effort typically required to provision and manage resources, enabling faster time-to-insight. As a result, organizations can run concurrent workloads without bottlenecks, allowing teams to focus on analysis and decision-making rather than managing infrastructure. This level of concurrency ensures high performance even as data volumes and user activity increase, all without impacting the end-user experience.

The data-sharing capabilities of Snowflake represent another significant benefit, particularly for organizations with multiple departments, partners, or external stakeholders. With traditional data warehousing, sharing data frequently involves creating copies or manually distributing files, leading to inefficiencies and version control issues. Snowflake enables direct, real-time data sharing without unnecessary duplication of data, allowing recipients to access the latest, accurate information like having it stored locally. Such robust data sharing facilitates collaboration and helps maintain consistency across different users or organizations, improving overall operational efficiency.

Snowflake's built-in security features are also a decisive advantage for enterprises concerned with data protection. It supports encryption in transit and at rest, ensuring that sensitive data remains secure throughout its lifecycle. Furthermore, Snowflake uses role-based access control (RBAC), enabling organizations to precisely define who can access specific data, tables, or query results. This granularity helps ensure that only authorized users have access to critical information, which is particularly important for organizations dealing with regulated or confidential data.

Finally, Snowflake provides advanced data governance capabilities essential for managing large-scale data environments. Features like automatic data backups, data replication across multiple regions, and failover protection ensure that data is always available, even during hardware failures or disasters. The platform also supports audit logs and monitoring tools that help track data access and usage, providing transparency and accountability. These features help businesses maintain compliance with industry regulations, improve data security, and ensure business continuity.

In summary, Snowflake offers many benefits, making it a robust, scalable, and secure solution for modern data warehousing. Its ability to handle diverse data types, provide elastic scalability, simplify data sharing, and offer advanced security and governance features positions it as a top choice for organizations seeking to manage and leverage their data efficiently. By combining these capabilities into a single, integrated platform, Snowflake helps organizations reduce operational complexity, optimize costs, and drive better insights from their data.

Snowflake's Multi-Cluster Shared Data Architecture

Snowflake's Multi-Cluster Shared Data Architecture is a key feature differentiating it from traditional data warehousing solutions at a fundamental level. At its core, the architecture design separates compute and storage, allowing them to scale independently. This separation means that Snowflake can elastically allocate resources for computing needs while maintaining a centralized storage layer. Data storage happens in a single location, but multiple virtual warehouses (compute clusters) can access this data simultaneously without interference. This architecture ensures that workloads do not impact each other, even when running concurrently, providing enhanced performance and efficiency.

A significant advantage of this architecture is its scalability. Since compute resources are decoupled from storage, businesses can scale each component based on their needs. If query volume or complex computation surges, additional compute clusters can be added dynamically without affecting the storage layer. Conversely, if data storage needs increase, the system can scale the storage without unnecessarily expanding compute resources. This flexibility allows organizations to optimize costs and performance based on their specific workload demands.

The multi-cluster nature of Snowflake's architecture also provides robust support for concurrent workloads. Organizations can run multiple workloads—such as queries, data loads, and transformations—without performance degradation, even if these workloads are resource-intensive. Each virtual warehouse operates independently, meaning long-running queries or heavy data processing tasks will not block or slow down other operations. This isolation ensures that users or teams working in parallel do not experience any interference, which is particularly valuable in environments where teams need to access the same data concurrently.

Another notable benefit of the multi-cluster architecture is its automatic scaling feature. Snowflake can automatically add or remove compute clusters in response to workload demands. For example, if a particular virtual warehouse reaches its resource limits, Snowflake can automatically scale up by adding additional clusters to handle the increased load. Once the demand subsides, the system can scale down to reduce costs. This level of automation ensures high availability and performance without requiring manual intervention or complex configuration, making Snowflake a highly efficient and cost-effective solution for organizations.

Snowflake's Unique Storage and Compute Separation

Snowflake's unique storage and compute separation is one of its most innovative features, offering a flexible, scalable architecture that addresses many of the challenges posed by traditional data warehouse systems. In traditional data platforms, storage and compute are tightly coupled, meaning that expanding compute resources often requires expanding storage, and vice versa. This can lead to inefficiencies and cost issues, especially when scaling up resources for specific workloads. Snowflake resolves this by keeping storage and compute resources completely independent of one another, allowing them to be scaled separately based on demand.

The storage layer in Snowflake is designed to handle vast amounts of data without impacting performance. It uses a centralized, cloud-based repository that stores data in a columnar format. This design optimizes query performance and enables Snowflake to process large volumes of data more efficiently. Because the storage is independent of compute, it can grow as needed without requiring additional compute resources, ensuring that organizations only pay for the storage they actually use. This separation also means that data is always available to any virtual warehouse, regardless of the compute resources allocated, enabling seamless access across multiple users and workloads.

On the compute side, Snowflake leverages virtual warehouses, which are essentially clusters of compute resources that can be allocated and deallocated as needed. Each virtual warehouse is completely isolated, meaning that workloads running on one warehouse do not affect others, even if they are working with the same data. This isolation is critical for managing workloads with varying resource demands. For example, a heavy ETL job running in one virtual warehouse won't interfere with a set of analytical queries being executed in another, ensuring consistent performance across the system.

Because storage and compute are decoupled, Snowflake offers significant cost optimization opportunities. Users can scale compute resources up or down based on workload demands without worrying about impacting storage costs. For example, a business can allocate additional compute power for high-priority tasks or heavy queries during peak times while reducing compute capacity during off-peak hours to save on costs. Similarly, organizations can keep their storage layer large enough to handle vast amounts of data without being forced to overprovision compute resources unnecessarily.

Lastly, this separation provides a high level of concurrency without performance degradation. Snowflake's architecture allows multiple users and teams to access and query the same data simultaneously, each with their own dedicated virtual warehouse. This means that one user's heavy computations or data transformations will not block or slow down another user's queries. As a result, Snowflake ensures high availability, responsive query performance, and the ability to handle complex, concurrent workloads, all while optimizing costs by scaling storage and compute independently.

Snowflake's Data Sharing Capabilities

Snowflake's data-sharing capabilities enable organizations to share data securely and efficiently without duplicating data. Unlike traditional data-sharing methods, which often involve copying data across multiple systems, Snowflake allows users to directly share data with other Snowflake accounts in a read-only format. This capability eliminates the need for time-consuming and error-prone data transfers, reducing redundancy and storage costs. The data remains in its original location, and Snowflake users can access it in real-time without disrupting the underlying data storage or performance.

One key advantage of Snowflake's data sharing is its real-time access to shared data. When an organization shares data with another Snowflake account, the receiving account gains access to live, up-to-date information. This live sharing of data ensures that stakeholders can always work with the most current version of the data, whether for reporting, analytics, or machine learning purposes. The ability to instantly share data across different departments or external partners facilitates collaboration and decision-making, ensuring everyone works with consistent and accurate information.

Another critical benefit of Snowflake is its secure data sharing. Unlike traditional data-sharing methods, which may involve transferring sensitive data over unsecured channels, Snowflake allows for precise control over access. Data owners can share specific tables or views with defined permissions, ensuring that sensitive information is only accessible by authorized parties. Additionally, Snowflake enables organizations to enforce data governance policies even when sharing data externally. This level of control helps organizations comply with data privacy regulations, such as GDPR or CCPA, while ensuring secure and compliant data-sharing practices.

Another powerful feature of Snowflake's data sharing is its ability to share data across different cloud platforms. Since Snowflake operates on top of major cloud providers like AWS, Google Cloud, and Azure, organizations can seamlessly share data across various environments, regardless of their cloud platform. This cross-cloud compatibility simplifies data sharing between organizations that may use different cloud infrastructures, breaking down barriers between silos and promoting collaboration between diverse stakeholders. Whether sharing data internally or externally, Snowflake's data-sharing capabilities streamline data access and foster deeper, more agile collaboration across organizations and platforms.

Performance Optimization in Snowflake

Performance optimization in Snowflake ensures that queries and workloads run efficiently while maintaining the flexibility to scale with changing business needs. One of the primary ways Snowflake optimizes performance is through its automatic scaling feature. Snowflake's architecture separates storage and compute, allowing compute resources to scale independently based on workload demands. This architecture design means that when there is a high volume of queries or processing jobs, Snowflake can automatically allocate more compute resources to handle the load, ensuring no impact on performance. When demand decreases, it can scale back to reduce unnecessary resource consumption, optimizing cost efficiency.

Another key aspect of Snowflake's performance optimization is its query optimization capabilities. Snowflake utilizes a sophisticated query execution engine that optimizes how it executes queries, ensuring that it handles operations like joins, aggregations, and filtering most efficiently. Snowflake also uses materialized views and caching to speed up repetitive queries. Frequently accessed data can be stored in memory, reducing the time needed for future retrievals and improving the overall speed of query execution. The engine also supports automatic query rewrites to ensure that queries are executed most efficiently without requiring manual intervention from the user.

In addition to query optimization, Snowflake offers tools like result caching and data clustering to enhance performance. Result caching stores the results of frequently executed queries in memory, so if the same query is rerun, Snowflake can return the cached result instantly, greatly reducing execution time. Meanwhile, clustering allows for grouping related data within large tables to reduce the time it takes to scan and filter the data. By organizing the data to make it easier to retrieve, clustering can significantly improve performance for large datasets with complex queries.

Snowflake's multi-cluster architecture also plays a key role in performance optimization. This feature enables Snowflake to deploy multiple compute clusters that can operate simultaneously, ensuring that workloads do not compete for resources. For example, one cluster can handle user queries while another can manage data loading or transformation tasks, preventing delays and bottlenecks. This separation of workloads ensures that performance remains optimal even during high-demand periods. Additionally, Snowflake can automatically increase the number of clusters based on workload volume, ensuring that the system always has the resources necessary to maintain high performance.

Snowflake for Data Security and Compliance

Snowflake provides robust data security features designed to meet modern organizations' stringent demands, particularly those in industries with sensitive data. One of the foundational aspects of Snowflake's security is its end-to-end encryption. All data, whether at rest or in transit, is encrypted using strong encryption protocols. This encryption ensures that sensitive information remains protected throughout its lifecycle, from data ingestion to storage and access. Additionally, Snowflake leverages role-based access control (RBAC), allowing administrators to assign permissions based on the user's role within the organization. This fine-grained access control ensures that only authorized individuals can view or modify data, helping mitigate risks related to unauthorized access.

In addition to data encryption and RBAC, Snowflake offers multi-factor authentication (MFA), a critical layer of security that ensures that only legitimate users can access the system. MFA is essential for organizations dealing with confidential or regulated data, as it adds an extra layer of protection beyond just usernames and passwords. Snowflake's security model also includes network security features, such as IP whitelisting and virtual private networks (VPNs), which allow organizations to restrict further and control the environments from which their Snowflake instance can be accessed. These combined security measures help ensure that Snowflake's cloud-based platform remains secure, even in highly regulated industries.

Snowflake meets various industry standards and regulatory requirements on the compliance front, making it a suitable solution for organizations in sectors like finance, healthcare, and government. Snowflake has obtained certifications for GDPR, HIPAA, SOC 1, SOC 2, and PCI DSS, among others. These certifications ensure that Snowflake adheres to the strictest security and compliance protocols, giving customers confidence that their data is being managed securely and in compliance with relevant laws. Snowflake's ability to manage data privacy and compliance at scale makes it an ideal choice for enterprises looking to leverage cloud data warehousing while meeting regulatory standards.

Integrating Snowflake with Third-Party Tools and Platforms

Snowflake's ability to integrate seamlessly with third-party tools and platforms is one of its key strengths, enabling organizations to leverage a vast ecosystem of solutions for data analytics, business intelligence, machine learning, and more. Snowflake offers a rich set of native connectors and APIs, allowing integration with a wide range of tools, from ETL platforms like Fivetran and Stitch to analytics tools like Tableau and Power BI. These integrations enable data engineers and analysts to quickly extract, load, and visualize data from Snowflake, making it easier to derive actionable insights hidden within the data. Additionally, Snowflake's support for ODBC and JDBC drivers facilitates seamless connections with various SQL-based applications, allowing users to interact with their data as if it were stored in a traditional database, without the need for complex data transformations.

For organizations leveraging machine learning and AI platforms, Snowflake provides compatibility with tools such as DataRobot, Azure Machine Learning, and Amazon SageMaker. By integrating Snowflake with these platforms, data scientists can perform advanced analytics, build predictive models, and train algorithms directly on the data stored in Snowflake, avoiding the need for data movement between disparate systems. Snowflake also supports Python and R through Snowpark, allowing for custom data processing, model training, and application development within the platform. The Snowpark integrated environment significantly reduces friction between data storage, processing, and machine learning workflows.

In addition to analytics and machine learning, Snowflake's integration capabilities extend to data sharing and collaboration with third-party tools. Snowflake's Data Marketplace allows organizations to share and exchange data with external partners, vendors, or customers, making accessing and utilizing datasets from other Snowflake users easier. The platform's ability to seamlessly integrate with data integration and workflow automation tools enhances its collaborative capabilities, allowing organizations to build end-to-end data pipelines and workflows across multiple platforms. Whether real-time data streaming with Apache Kafka and Apache Spark or orchestrating data pipelines with tools like Airflow, Snowflake offers the flexibility and scalability needed to integrate with modern data ecosystems.

Use Cases and Industry Applications of Snowflake

Snowflake's flexibility and scalability make it an ideal solution for various industries, each with unique data needs. One of the key use cases for Snowflake is in financial services, where organizations need to store, analyze, and protect massive volumes of financial data. Snowflake's secure data-sharing capabilities allow financial institutions to collaborate seamlessly with external partners, auditors, and regulators while maintaining strict data security and compliance standards. The platform's ability to handle large-scale analytics and its support for real-time data streaming enables businesses to gain actionable insights into market trends, financial performance, and risk management. This makes Snowflake an invaluable tool for fraud detection, credit risk assessment, and regulatory reporting.

Snowflake is increasingly used in the healthcare industry to support data-driven decision-making and improve patient care. Healthcare organizations collect vast amounts of data from electronic health records (EHR), medical devices, wearables, and other sources, often in various formats and systems. Snowflake's multi-cloud capabilities and data integration tools enable healthcare providers to centralize this data, making it easier to analyze and derive insights for better patient outcomes. Additionally, the platform's ability to provide data governance ensures patient privacy and compliance with regulations like HIPAA. With the ability to scale quickly and provide real-time insights, Snowflake is becoming an essential platform for predictive analytics in healthcare, helping organizations improve diagnosis accuracy, personalize treatment plans, and optimize operational efficiency.

For e-commerce and retail companies, Snowflake is critical in driving customer personalization and optimizing inventory management. By integrating data from customer interactions, online transactions, and supply chains, Snowflake allows businesses to create detailed customer profiles and segmentations. This data-driven approach enables targeted marketing campaigns, dynamic pricing strategies, and personalized recommendations that enhance customer satisfaction and boost revenue. Snowflake's performance optimization features, such as automatic scaling, ensure that businesses can handle peak demand during seasonal sales or promotional events without compromising query performance or user experience. This scalability is crucial for e-commerce platforms that experience unpredictable traffic and must maintain high availability.

In the media and entertainment industry, Snowflake is helping organizations manage and analyze large datasets generated by content distribution, viewership metrics, and customer interactions. Media companies use Snowflake to centralize data from multiple sources, including digital platforms, TV networks, and social media channels, enabling them to track audience behavior, identify trends, and optimize content strategies. By combining data from multiple platforms in a secure environment, Snowflake provides an efficient solution for audience segmentation, content recommendation engines, and targeted advertising. Furthermore, the platform's high-performance compute capabilities ensure that companies can process large amounts of unstructured data, such as video and audio files, to deliver real-time insights and support faster content delivery.

Snowflake monitors and optimizes production processes, supply chain operations, and predictive maintenance efforts for manufacturing industries. Manufacturing companies are increasingly adopting IoT sensors and other connected devices to track the performance of machinery and gather real-time data on production lines. Snowflake's ability to handle large-scale data streams and integrate with IoT platforms makes it ideal for storing and analyzing this data. By leveraging Snowflake's real-time data processing capabilities, manufacturers can identify inefficiencies, predict equipment failures, and optimize inventory management. This enables them to reduce downtime, improve product quality, and streamline their supply chains, all while minimizing costs and maximizing profitability.

Snowflake is transforming how companies manage and utilize network data in the telecommunications industry. Telecom providers collect vast amounts of data on network performance, customer usage patterns, and device interactions, which can be challenging to analyze due to the volume and complexity of the data. Snowflake's cloud-native architecture allows telecommunications companies to consolidate data from multiple sources, providing a unified view of network performance and customer behavior. Using Snowflake for real-time analytics, telecom companies can optimize their networks, improve customer experiences, and identify opportunities for new service offerings. Additionally, Snowflake's ability to scale quickly ensures that telecom providers can handle peak data loads during high-demand periods, such as when launching new products or services.

Snowflake is being leveraged in the education sector to centralize student data, track academic performance, and improve learning outcomes. Educational institutions often deal with data scattered across multiple systems, including learning management systems (LMS), student information systems (SIS), and administrative platforms. Snowflake's ability to integrate disparate datasets allows schools and universities to view student performance, attendance, and engagement comprehensively. By applying advanced analytics to this data, educational institutions can identify at-risk students, improve course offerings, and personalize learning experiences. Snowflake's data-sharing capabilities also enable collaboration between schools, government agencies, and research institutions, fostering innovation and improving educational outcomes on a broader scale.

Snowflake also impacts energy and utilities, where companies use it to optimize resource management, monitor consumption patterns, and improve sustainability efforts. Energy providers collect data from various sources, including smart meters, sensors, and weather forecasts, all of which must be analyzed for better decision-making. Snowflake's ability to store and analyze large volumes of time-series data makes it a powerful tool for tracking energy usage, detecting anomalies, and predicting demand fluctuations. By integrating data across different energy grids, Snowflake enables companies to optimize distribution, reduce waste, and improve the reliability of their services. Additionally, Snowflake's support for machine learning models allows energy companies to forecast future demand and optimize energy production, helping them transition to more sustainable and efficient operations.

Finally, Snowflake is increasingly being used for data transparency, citizen engagement, and fraud detection in government and public sector organizations. Governments collect and store vast amounts of data across various departments, from healthcare and social services to law enforcement and transportation. Snowflake's ability to centralize and securely share this data enables government agencies to collaborate more effectively and make data-driven policy decisions. Public agencies can also use Snowflake to improve public safety by analyzing crime data, predicting future criminal activities, and deploying resources more efficiently. Snowflake's security features and compliance certifications, such as SOC 2 and GDPR, ensure that sensitive public sector data is protected while remaining accessible for authorized analysis and reporting.

Conclusion

Snowflake's versatility and scalable architecture have established it as a powerful tool across various industries, transforming how organizations manage, analyze, and share data. From finance to healthcare, retail, and beyond, Snowflake's ability to handle vast amounts of diverse data while providing real-time insights and ensuring security has made it a go-to platform for businesses seeking to drive data-driven decision-making and innovation. The platform's ability to seamlessly integrate with a wide variety of data sources and third-party tools, combined with its multi-cloud capabilities, positions it as a key enabler for organizations looking to remain competitive in today's data-driven world.

As industries continue to evolve and embrace digital transformation, Snowflake's use cases will only expand, with companies increasingly relying on it to solve complex data challenges. Whether optimizing operations, enhancing customer experiences, or gaining insights into new business opportunities, Snowflake offers a comprehensive solution that scales to meet the needs of modern enterprises. Its industry-specific applications demonstrate its adaptability, and its future in reshaping data management and analytics remains bright as more organizations harness its power for their business goals.

CHAPTER 6

Introduction to Git

Version control is a critical aspect of modern software development, and Git has become the go-to tool for managing code changes in a collaborative environment. Whether working on individual projects or as part of a team, Git provides a robust system for tracking changes, managing different versions of code, and ensuring that collaboration runs smoothly. With its distributed nature, Git allows developers to work on code independently, merge their changes seamlessly, and maintain a complete history of every modification. This chapter will introduce the core concepts and fundamental workflows of Git, helping new and experienced developers understand how to use it effectively.

As software development becomes increasingly collaborative, especially with distributed teams and open-source projects, Git's role in facilitating smooth collaboration and efficient project management has become undeniable. Understanding how to leverage Git for version control is essential for maintaining code integrity, ensuring proper project documentation, and managing various stages of development. This chapter will explore Git's architecture, basic commands, and key workflows, laying the foundation for mastering version control and integrating it into your development process.

What Is Version Control?

Version control is a system that helps track changes to files and projects over time. It allows multiple people to work on the same project simultaneously, recording each modification made. Whether a software project, document, or data set, version control ensures that each change is logged and can be reviewed or reverted if necessary. This makes it easier to collaborate with others, maintain a history of changes, and ensure that the most current version of a file is always accessible.

R. L. Steelman, Jr., *Mastering Snowflake DataOps with DataOps.live*, https://doi.org/10.1007/979-8-8688-1754-0_6

The core idea behind version control is to store a history of changes, including details about who made the change and why. Knowing the "why" can be especially useful in collaborative environments where multiple team members track modifications in a structured way. Version control systems prevent conflicts and minimize the risk of overwriting someone else's work. Git saves each version or snapshot of the project as a "commit," and these commits can be reviewed, reverted, or merged with other changes, ensuring the project's integrity and consistency.

There are two primary types of version control: centralized and distributed. A centralized version control system (CVCS) has a single central repository where all the files and their changes are stored. Users check out files from this repository, make changes, and then commit them back. A distributed version control system (DVCS), such as Git, allows users to have a complete copy of the repository, including its history, on their personal machine. This allows users to work offline and commit changes without relying on a central server.

Version control is especially valuable for tracking code changes in software development, where numerous project iterations are common. It helps developers avoid issues such as conflicting code updates or the accidental loss of significant code changes. By enabling developers to work on different branches, version control systems allow them to experiment with new features or bug fixes without affecting the main project. If the changes are successful, they can be merged back into the main branch; if not, they can be discarded without any permanent consequences.

In addition to supporting collaboration and reducing the risks of errors, version control promotes a culture of accountability. With a clear record of who made each change, it's easy to trace any issues back to their source. This transparency fosters better teamwork and communication, as team members can see exactly what others are working on and understand the rationale behind specific changes. Version control ensures that development processes are organized, efficient, and transparent, making it an essential tool for modern software development and other collaborative projects.

Why Is Version Control Important?

Version control is crucial because it provides a structured way to manage changes, ensuring every modification is tracked and reversible. Managing changes in this fashion helps prevent the loss of essential data or code and allows teams to maintain a complete history of changes, making it easy to review or undo previous work if necessary. In a collaborative environment where multiple individuals may contribute to the same project, version control helps avoid conflicts and ensures that everyone's contributions are accurately recorded. Without version control, it would be easier to overwrite or lose valuable work, leading to clarity and errors.

Another key reason version control is important is its ability to enhance collaboration. When multiple developers or team members work on a project simultaneously, version control allows them to work independently on different parts of the project without interfering with each other's work. By isolating changes in different branches, and once they are tested and verified, users can merge them into the main project. This flexibility not only boosts productivity but also promotes innovation by allowing developers to experiment without worrying about disrupting the core functionality of the project.

Version control also enables better project tracking and accountability. Every change made to the project is part of the logging process, which includes information about who made the change, when it was made, and why it was necessary. This record of changes provides valuable insight into the project's evolution and helps identify any issues or bugs that might arise. By maintaining this historical record, teams can track progress, measure productivity, and ensure that decisions are documented, which fosters transparency and reduces misunderstandings.

Lastly, version control is essential for ensuring a project's stability and reliability, especially when dealing with complex or large codebases. As projects grow, it becomes more difficult to manage changes manually. Version control systems enable isolating, testing, and integrating changes systematically, reducing the risk of introducing bugs or errors. By giving developers the tools to roll back to previous versions, compare different versions, and merge code efficiently, version control ensures that projects remain functional and stable over time, even as they evolve.

What Is Git?

Git is a distributed version control system that allows developers to track changes in their codebase over time. Unlike traditional version control systems, where users change artifacts in a central repository, Git enables every developer to have a local copy of the entire repository, including its history. This decentralized approach provides several advantages, such as the ability to work offline, collaborate without needing constant connectivity to a central server, and easily manage and merge changes from multiple sources. Git's flexibility allows developers to work on different features or bug fixes simultaneously in isolated branches and later combine these changes through a process called merging.

At its core, Git relies on snapshots rather than just tracking differences between files. Each time a developer commits a change, Git stores a snapshot of the project at that specific point in time, making it easy to revert to any previous version or analyze the project's history. This snapshot-based approach is efficient and fast, enabling developers to quickly switch between different project versions. Git's powerful branching and merging features also allow teams to work on new features or fixes without interrupting the main codebase, which is essential for maintaining stability in large projects.

Git is commonly used alongside platforms like GitHub, GitLab, and Bitbucket, which provide cloud-based hosting for Git repositories. These platforms offer additional tools for collaboration, such as pull requests, issue tracking, and project management features. By integrating Git with these services, developers can streamline workflows, track bugs, and collaborate seamlessly on complex software projects. Git's widespread adoption across the software development industry makes it a foundational tool for version control, ensuring developers can work more efficiently, reliably, and collaboratively on projects of all sizes.

Understanding Git's Core Principles and Workflow

Git has built its core principles around several key concepts that make it a powerful version control system. At its heart, Git operates using a distributed model, which means every user has a personal local copy of the repository, including its entire history. This distributed nature allows developers to work independently without needing constant access to a central server. The local repository enables users to make changes, commit those changes, and create new branches, all while disconnected from the network. Once the developer is ready, they can sync their changes with the remote repository, where the work of multiple team members can be integrated.

A fundamental concept in Git is the commit. Each commit represents a snapshot of the project at a specific point in time. This snapshot records changes made to the codebase, including adding, deleting, or modifying files. A commit is associated with a commit message, which provides context for the changes. This history of commits forms a linked chain, allowing developers to track the evolution of the project. Reverting to any previous commit is straightforward in Git, enabling teams to undo mistakes, view the progression of the code, or explore how a feature was implemented over time.

Branches are another essential principle in Git. They allow for independent lines of development, where different features or bug fixes can be worked on without affecting the main codebase. Branching enables developers to experiment with new ideas without disrupting the stability of the project. Once a feature is complete, the branch should be merged back into the core branch, often called "master" or "main." Git handles the complexities of merging branches, ensuring that code from different contributors can be integrated smoothly. Conflicts may arise when changes overlap, but Git provides tools for resolving these conflicts during the merge process.

Git's workflow is designed to be flexible and supports several models, depending on the needs of the project or team. The most common workflow is the "feature branch" workflow, where developers create a new branch for each feature or bug fix they are working on. Once the feature is complete and tested, it is merged back into the main branch. For teams that require more complex workflows, Git also supports the "GitFlow" model, which introduces multiple long-lived branches such as development for ongoing work and releases for preparing new software versions. In this case, Git allows for a more structured release management process, ensuring each branch serves a specific purpose.

Another key component of Git's workflow is the pull request (or merge request). When a developer is ready to merge their changes into a shared branch, they create a pull request, which notifies the team that the changes are ready for review. This is an opportunity for other developers to review the changes, discuss potential improvements, and ensure that the code meets the project's standards before it is merged into the main codebase. Pull requests also help maintain the project's integrity, as they act as a form of peer review, ensuring that only high-quality code gets integrated.

Git's workflow emphasizes collaboration, efficiency, and control. Developers work independently in their local repositories and branches, which reduces the risk of interfering with each other's work. The workflow allows teams to implement feature changes in parallel, merge code efficiently, and ensure that the project's main branch remains stable. Using Git's core principles, such as commits, branches, and pull requests, teams can maintain a transparent, organized development process, even in large, complex projects with multiple contributors. This structure is essential for managing code in an agile, collaborative environment.

Best Practices for Using Git in Collaborative Development

Using Git effectively in collaborative development requires adherence to certain best practices that promote teamwork, ensure code quality, and prevent conflicts. One of the foundational practices is maintaining a clear and consistent branching strategy. Teams should agree on a branching model that suits their workflow, such as GitFlow, trunk-based development, or a feature-branch workflow. For example, developers can work on individual features or bug fixes in isolated branches and only merge into the main branch once the work is complete and tested. This isolation helps protect the main branch from unstable code and ensures a clear structure in the repository.

Commit messages play a crucial role in collaborative development. Every commit should have a meaningful and descriptive message explaining the changes' purpose. Rather than using vague phrases like "fixed bugs" or "updates," a well-written commit message might say, "Fixed null pointer exception in user authentication flow" or "Refactored login service to improve performance." Descriptive commit messages provide context for team members and make understanding the project's history easier. These commit messages are constructive during code reviews, debugging, or revisiting the repository months or years later.

Another essential practice is regularly syncing with the remote repository. Developers should frequently pull the latest changes from the shared branches to ensure their local branches are current. These reviews minimize the risk of encountering significant merge conflicts when pushing changes back to the shared repository. Frequent synchronization fosters a smoother integration process and helps developers stay aligned with ongoing work. Additionally, resolving minor conflicts early is much easier than dealing with significant, accumulated conflicts at the end of a sprint or development cycle.

Code reviews are a critical aspect of collaborative Git usage. Before merging a feature branch into the main branch, developers should submit their changes for peer review, often using pull requests. This step not only ensures the quality of the code but also fosters knowledge sharing within the team. Reviewers can identify potential issues, suggest improvements, and learn from each other's work. Additionally, conducting reviews helps enforce coding standards and ensures all contributions align with the project's guidelines. Peer reviews can also act as a checkpoint to catch mistakes that may have gone unnoticed during development.

Using Git tags and versioning is another best practice that enhances organization and tracking. Tags are beneficial for marking specific points in the project's history, such as releases or milestones. Semantic versioning, for instance, allows teams to communicate the significance of changes in a release (e.g., significant updates, minor enhancements, or patches). By tagging these milestones, teams can easily roll back to previous versions if issues arise, and software users can understand the nature of changes between versions.

Lastly, setting up automated testing and continuous integration (CI) pipelines is a key practice for maintaining stability in collaborative environments. Automated tests should run on every pull request to ensure new changes do not break existing functionality. These tests provide immediate feedback to developers, helping them address issues before merging their changes. Integration pipelines can also automate tasks like building the application, running security scans, and deploying code to staging environments. By integrating Git with CI/CD tools, teams can maintain a high-quality codebase while accelerating development workflows. When consistently applied, these practices make Git a powerful tool for collaborative software development.

Conclusion

Version control, and specifically Git, serves as a cornerstone of modern software development, enabling teams to collaborate efficiently and maintain a high-quality codebase. By understanding Git's core principles and implementing best practices, teams can streamline workflows, prevent costly errors, and enhance overall productivity. Whether it's managing changes across branches, ensuring robust documentation through commit messages, or facilitating seamless collaboration with pull requests, Git provides the tools necessary to handle the complexities of development projects. Its versatility makes it suitable for both small teams and large-scale, distributed organizations.

As developers continue to adopt Git in their workflows, the importance of mastering its features and practices cannot be overstated. The ability to efficiently track changes, resolve conflicts, and integrate continuous testing and deployment processes is invaluable in today's fast-paced development environments. By embracing Git as a tool and a discipline, teams can build more resilient, scalable, and innovative software, laying the foundation for success in an increasingly competitive industry.

CHAPTER 7

Getting Started with Git

Git is a powerful tool that underpins modern version control practices, making it an essential skill for developers at all experience levels. While its vast array of features may seem daunting to beginners, mastering the basics of Git unlocks its full potential as a collaboration and code management tool. From tracking changes to managing branches and resolving conflicts, Git empowers developers to maintain a clean, organized, well-documented codebase, even in complex projects with multiple contributors.

This chapter serves as a practical guide for getting started with Git. Whether you're new to version control or transitioning from another system, we'll walk through the foundational steps to use Git effectively. From setting up your first repository to committing changes and exploring basic commands, this chapter provides the knowledge you need to integrate Git into your development workflow confidently. By the end, you'll have a solid foundation to build upon as you delve deeper into more advanced Git capabilities.

Basic Git Concepts

Git is built upon a foundation of concepts, making it a robust and efficient version control system. At its core, Git is a distributed version control system, meaning every user has a complete copy of the repository, including its full history. This approach ensures that developers can work independently without relying on a central server, enhancing collaboration and reliability. Should a server failure occur, any developer's local repository can serve as a backup to restore the project.

A Git repository is the central structure where all files and their revision histories are stored. Repositories can exist locally on a developer's machine or remotely on platforms like GitHub, GitLab, or Bitbucket. Developers can track changes within a repository across different states: working directory, staging area, and commit history. These states are integral to how Git handles changes and provide a structured process for managing updates to a codebase.

R. L. Steelman, Jr., *Mastering Snowflake DataOps with DataOps.live*, https://doi.org/10.1007/979-8-8688-1754-0_7

One of Git's most important concepts is committing changes. A commit represents a snapshot of the code at a particular point in time. Developers use commits to save and document changes, making it easier to track a project's evolution. Git accompanies each commit with a unique hash ID and a merge, allowing developers to quickly make specific changes and document granular and meaningful commits. This process will enable teams to maintain an organized project history.

Branches are another fundamental feature of Git. They allow developers to work on isolated development lines without affecting the main codebase. For example, creating a new branch for a feature enables experimentation and iteration without risking disruptions to the stable version of the project. Once the feature is complete and tested, it can be merged back into the main branch. This branching model supports parallel development and reduces the likelihood of conflicts.

Merging is the process of integrating changes from one branch into another. Git provides tools to handle merges effectively, even in complex scenarios. While simple merges occur automatically, conflicts may arise when two developers modify the same file and make the file incompatible. Git flags these conflicts, enabling developers to resolve them manually. This ensures that the final code integrates the best elements from all contributors.

The concept of staging is another key aspect of Git's workflow. Before Git commits changes, it adds those changes to the staging area. This step allows developers to review and group related changes before saving them as a single commit. The staging area acts as a buffer, offering control over what to include in each commit. This granular approach to committing changes makes it easier to track and revert specific updates if necessary.

Remote repositories facilitate collaboration by serving as centralized locations where developers can push their changes or pull updates made by others. Commands like git push and git pull synchronize local and remote repositories, ensuring all contributors work with the most up-to-date code. This system fosters effective teamwork, even when developers are distributed across different locations.

Git also employs a concept called tags, which mark specific commits as significant milestones, such as release versions. Unlike branches, tags are static and provide a snapshot of the repository at a given point. They are especially useful for tracking stable releases or deployment-ready versions of a project.

Another core concept in Git is rebasing, which allows developers to integrate changes by reapplying commits on top of a different base. While rebasing can streamline a project's commit history, it requires careful handling to avoid introducing inconsistencies. Developers must understand when to use rebasing versus merging to maintain the integrity of the repository.

Finally, Git's distributed nature provides unparalleled flexibility and resilience. Each local repository is a complete project version, enabling offline work and robust backups. With its powerful branching, staging, and merging capabilities, Git provides a comprehensive solution for managing code changes in projects of any size or complexity. Understanding these basic concepts sets the stage for more advanced workflows and ensures a smoother development process.

Installing Git

Getting started with Git requires installing it on your local machine (Figure 7-1). Git is compatible with multiple operating systems, including Windows, macOS, and Linux. Each system has its unique steps, but the overall process is straightforward and well-documented. The first step is to ensure you have administrative privileges on your computer, which is often required to install software. Once you have these permissions, you can download and install Git from the official Git website or use a package manager compatible with your operating system.

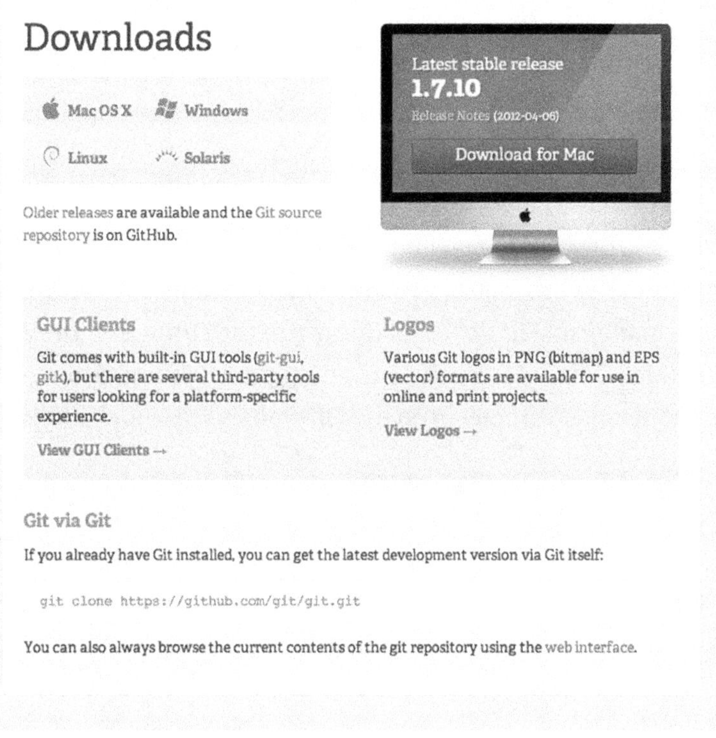

***Figure 7-1.** Git Homepage*

For Windows users, installing Git typically involves downloading the .exe file from the official Git website and running the installer. During installation, the setup wizard provides options to configure settings, such as the default text editor, the initial branch name, and how Git integrates with the command line. It's important to review these options carefully to tailor Git to your preferences (Figure 7-2). After completing the installation, you can verify it by opening the Command Prompt or Git Bash and running the command ***git --version***.

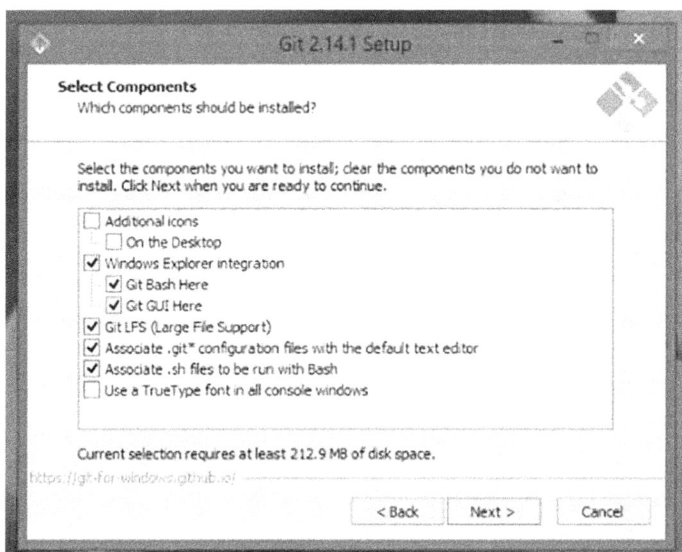

Figure 7-2. *Git Installer*

On macOS, Git is often pre-installed, especially on newer systems. However, if it isn't available or you want the latest version, you can install Git using Homebrew, a popular package manager (Figure 7-3). Simply run the command **brew install git** in your terminal. After installation, confirm it by running **git --version**. Alternatively, you can download the Git installer directly from the Git website if Homebrew is not an option.

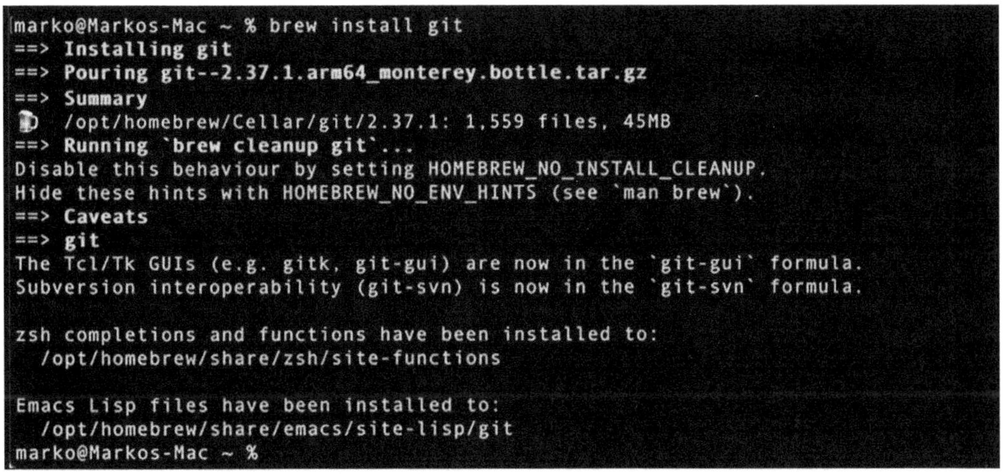

```
marko@Markos-Mac ~ % brew install git
==> Installing git
==> Pouring git--2.37.1.arm64_monterey.bottle.tar.gz
==> Summary
🍺  /opt/homebrew/Cellar/git/2.37.1: 1,559 files, 45MB
==> Running `brew cleanup git`...
Disable this behaviour by setting HOMEBREW_NO_INSTALL_CLEANUP.
Hide these hints with HOMEBREW_NO_ENV_HINTS (see `man brew`).
==> Caveats
==> git
The Tcl/Tk GUIs (e.g. gitk, git-gui) are now in the `git-gui` formula.
Subversion interoperability (git-svn) is now in the `git-svn` formula.

zsh completions and functions have been installed to:
  /opt/homebrew/share/zsh/site-functions

Emacs Lisp files have been installed to:
  /opt/homebrew/share/emacs/site-lisp/git
marko@Markos-Mac ~ %
```

Figure 7-3. *Git Installer: Homebrew*

Linux users typically install Git via their distribution's package manager. For example, on Debian-based systems like Ubuntu, the command **sudo apt-get install git** can be used. Red Hat-based systems, such as Fedora or CentOS, rely on the *yum* or *dnf* commands, such as **sudo yum install git.** It's important to update your package manager's repositories before installation to ensure you download the latest stable version. Again, you can verify the installation by running **git --version** in the terminal (Figure 7-4).

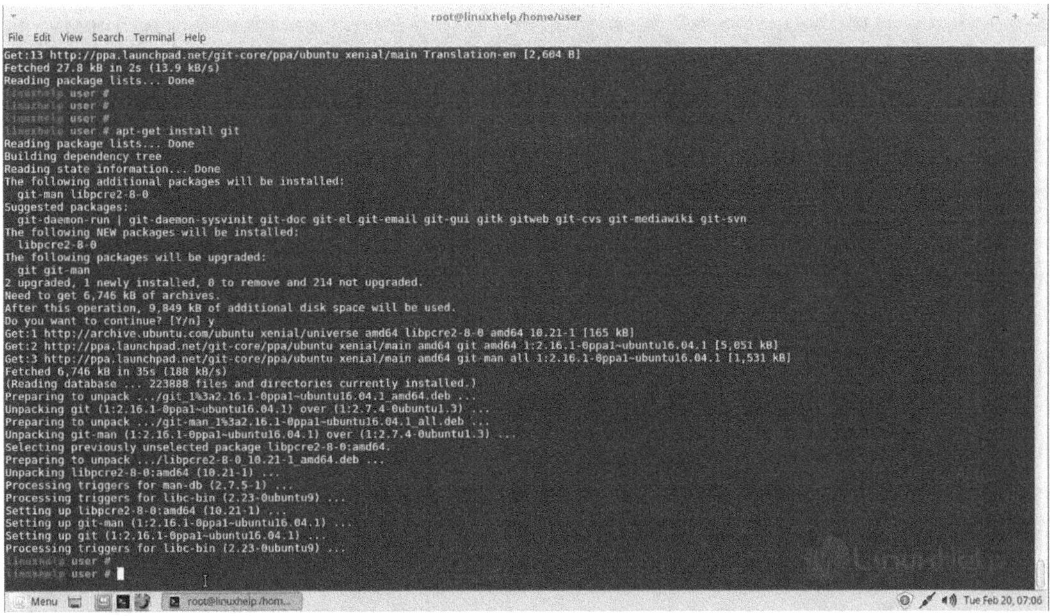

Figure 7-4. *Git Installer: Linux*

After installing Git, it's crucial to configure it properly before using it. Configuration involves setting your identity, which Git uses to label your commits. This is achieved by running two commands in the terminal: ***git config --global user.name "Your Name"*** and ***git config --global user.email "youremail@example.com"***. These settings are stored globally and apply to all repositories on your machine.

Another essential part of configuring Git is setting your preferred text editor for commit messages. By default, Git may use an editor like Vim, but you can specify another editor, such as Visual Studio Code or Nano. For instance, to set Visual Studio Code as your default editor, you would run ***git config --global core.editor "code --wait"***. This customization ensures a more comfortable experience when writing commit messages.

Some users prefer graphical user interfaces (GUIs) over command-line interactions. Git offers several GUI clients, such as GitKraken, SourceTree, and GitHub Desktop, which make Git easier to use, especially for beginners. These tools simplify tasks like committing, branching, and merging while visually representing your repository's structure. GUI tools can be downloaded separately, often with their installation steps, but they integrate seamlessly with the Git environment.

Once Git is installed and configured, you should familiarize yourself with its documentation and help commands. The git help command provides an overview of Git's capabilities, while specific commands like git help commit give detailed usage instructions for individual tasks. Access to Git's help system ensures you can quickly resolve any questions or issues as you learn.

In environments with strict security policies, Git installation might require additional steps, such as obtaining admin approval or setting up secure SSH keys for remote repository access. SSH keys are essential for secure communication with platforms like GitHub or GitLab. You can generate an SSH key pair using the command ssh-keygen and add the public key to your online Git account for seamless authentication.

Installing Git is a critical first step in mastering version control and collaborative development. By understanding the installation process for different operating systems and configuring Git to suit your preferences, you create a foundation for efficient and effective code management. Whether you're working alone or as part of a team, a well-installed and configured Git environment sets the stage for success in any development workflow.

Working with Git

Git is a powerful version control system that helps developers manage their codebases efficiently. Working with Git involves several core actions, including creating repositories, tracking changes, and managing branches. A Git repository is the fundamental structure where all project files and their change histories are stored. To create a repository, you can initialize a new one using git init, which creates the necessary files and folders for Git to function. Alternatively, you can clone an existing repository from a remote source like GitHub using git clone <repository-url>. These commands begin a version-controlled project, enabling teams to collaborate effectively.

Tracking changes in files is a critical part of working with Git. Once a repository is created, any changes to files within it must be staged and committed. The git add command stages changes by moving them to the staging area, while git commit records these changes in the repository's history. Each commit is accompanied by a message that describes what was changed, making it easier to understand the project's evolution over time. Creating meaningful and descriptive commit messages is important to facilitate clear communication among team members.

Branches are another essential feature in Git, allowing developers to simultaneously work on separate project versions. By default, Git initializes a repository with a main branch, but new branches can be created for specific tasks or features using git branch <branch-name>. Developers can switch between branches with git checkout or the more modern git switch. Once a task is complete, branches can be merged back into the main branch with git merge, consolidating the changes. This workflow isolates new development efforts, ensuring the main branch remains stable.

Conflict resolution is inevitable in working with Git, especially when multiple team members modify the same file. Conflicts occur when Git cannot automatically reconcile differences during a merge. Developers must manually resolve these conflicts by editing the affected files and marking them resolved with git add. After resolving conflicts, completing the merge with a git commit ensures that all changes are integrated. While conflicts can be challenging, Git provides tools and clear markers within files to simplify resolution.

Remote repositories are pivotal in Git workflows, enabling teams to collaborate across locations. Commands like git push and git pull synchronize changes between local and remote repositories. Pushing sends local commits to a remote server, while pulling retrieves the latest changes from the remote repository. Keeping the local repository updated ensures a consistent codebase and minimizes potential conflicts.

Popular platforms like GitHub, GitLab, and Bitbucket enhance Git's capabilities by providing web-based collaboration tools, issue tracking, and integration with continuous integration/continuous deployment (CI/CD) pipelines.

Effective use of Git also involves leveraging its history and log tools to track changes and understand the repository's state. The git log command displays the commit history, while git diff highlights specific changes between commits or branches. These tools are invaluable for debugging, auditing, and reviewing code changes. By mastering these and other Git commands, developers can work more efficiently, confidently manage their codebase, and foster seamless collaboration in any software project.

Collaboration with Git

Git fosters collaboration, making it a vital tool for teams working on software projects. By providing a decentralized system, Git allows team members to work on their individual copies of a repository, ensuring everyone can contribute independently without overwriting others' work. This type of personal work enables independence through branching, where team members can create their own branches to develop features or fix bugs. Once changes are ready, these branches can be merged into the main branch or other integration branches, maintaining a structured workflow. This model promotes parallel development while reducing the risk of conflicts and downtime in the main codebase.

Remote repositories are the backbone of Git's collaborative features. Platforms like GitHub, GitLab, and Bitbucket provide centralized hubs where teams can store, share, and synchronize their repositories. Using commands like git push and git pull, developers can share their updates with others or fetch the latest changes from the team. Collaboration becomes seamless as every team member works on the most current version of the code. These platforms also enable pull requests (or merge requests), which are structured ways to propose changes, review code, and discuss improvements before merging them into the main branch.

Collaboration with Git also benefits from its robust conflict resolution tools. When multiple developers edit the same file, Git highlights the differences and flags conflicts that need manual intervention. The ability to compare and act upon differences ensures that no changes are lost during the merging process. Clear communication and disciplined workflows, such as frequent pulls and timely pushes, help minimize conflicts. Teams often adopt conventions like enforcing code reviews or tagging releases to maintain consistency and quality in their shared codebase. Git's logs and diff tools further enhance collaboration by making every change traceable, which is invaluable for accountability and auditing.

Beyond code management, Git facilitates collaboration on documentation, configuration files, and even non-software projects. Its versatility extends to integrating with external tools for issue tracking, continuous integration (CI), and project management, creating a cohesive ecosystem for team collaboration. By adopting Git, teams gain a shared foundation for managing their work, improving communication, and fostering a culture of collaboration that can scale with the complexity of any project.

Aliases

Git aliases are a powerful feature that allows developers to create custom shortcuts for frequently used or complex Git commands. Instead of typing out long or repetitive commands, aliases enable you to use concise, memorable keywords to execute them. This customization can significantly speed up your workflow and reduce the cognitive load of remembering verbose commands. For instance, instead of typing git status, you could create an alias like gst, streamlining everyday tasks and improving efficiency.

Setting up Git aliases is straightforward and involves modifying the Git configuration file, typically done via the git config command. For example, to create an alias for git log --oneline, you could run ***git config --global alias.lg "log --oneline"***. Once set, typing ***git lg*** in the terminal will execute the configured command. Aliases can be as simple or complex as needed, covering everything from short commands like co for checkout to intricate ones that include multiple flags or options. These aliases are stored in your global or local .gitconfig file, making them easy to review and manage.

The use of aliases enhances productivity and enables teams to standardize workflows. By sharing a consistent set of aliases across a team, everyone can use the same shortcuts, reducing the risk of errors and ensuring that best practices are followed. For instance, a team might agree on a common alias for reviewing logs or staging files, ensuring uniformity in how tasks are performed. These shared aliases can be included in onboarding documentation or configuration scripts, making it easier for new team members to adapt quickly.

Despite their benefits, it's essential to use Git aliases judiciously. Over-reliance on aliases can make your workflow less transparent to others unfamiliar with your custom shortcuts. Balancing convenience and clarity is a good practice, as it keeps aliases intuitive and well-documented. For collaborative environments, sharing the alias configurations and ensuring they complement rather than obscure standard Git practices is crucial. With thoughtful implementation, Git aliases can be a powerful tool for both individual efficiency and team cohesion.

Tagging

Git tagging is a feature that allows you to create identifiable markers in your repository's history, often used to signify crucial events such as releases or milestones. Unlike branches, tags are immutable, meaning they don't change over time, providing a fixed reference point. Tags are particularly useful in software development workflows where specific commits, such as those corresponding to stable releases, must be highlighted for future reference. For instance, tagging a commit with v1.0.0 makes it easy to locate and work with that exact version of your codebase.

Git supports two main types of tags: lightweight and annotated. Lightweight tags are pointers to a specific commit and don't include additional metadata. They're quick to create and suitable for simple tasks without extra information. On the other hand, annotated tags are stored as full objects in the Git database and can include a message, the tagger's name, and a timestamp. Annotated tags are more robust and are generally recommended for marking releases or other significant events, as they carry more context.

Creating tags in Git is straightforward. For a lightweight tag, you can use the command ***git tag <tagname>***. To create an annotated tag, use ***git tag -a <tagname> -m "Tag message"***. These tags can later be pushed to a remote repository using ***git push origin <tagname>*** for individual tags or ***git push origin --tags*** to push all tags simultaneously. Retrieving tags is equally simple; commands like ***git show <tagname>*** or ***git checkout <tagname>*** allow you to view or work with the specific commit associated with a tag. Additionally, tags can be listed with *git tag* for an overview of all available tags in the repository.

Tagging enhances project organization and release management by making differentiating between development and production-ready versions easy. When combined with other Git features like branching, tagging enables developers to manage version control efficiently, ensuring a clear distinction between experimental changes and stable code. Whether for internal development purposes or public releases, Git tags provide a reliable way to pinpoint and reference specific commits, adding clarity and structure to your repository's history.

Remotes

Git remotes are references to external repositories that facilitate collaboration and centralized code management, typically hosted on platforms like GitHub, GitLab, or Bitbucket. A remote acts as a link between your local repository and its counterpart on a server. These remotes are integral to distributed version control systems, allowing team members to share changes, synchronize their work, and collaborate efficiently, even when working from different locations.

The most commonly used Git remote is named origin, typically created when a repository is cloned or initialized with a connection to a remote server. The origin remote points to the URL of the external repository and serves as the default target for commands like git push and git pull. Developers can inspect their remotes with the command ***git remote -v***, which displays all remote connections and their respective URLs. It's also possible to add, rename, or remove remotes using commands like ***git remote add <name> <url>*** or ***git remote remove <name>***.

Using remotes efficiently involves some key workflows, such as pulling updates from the remote repository, pushing local changes, and resolving conflicts. The git pull command fetches changes from the remote and merges them into your local branch, while git push uploads your changes to the remote. These operations ensure that your repository stays in sync with the team. For more granular control, the git fetch command retrieves changes without merging them, allowing you to review updates before integrating them into your local branch.

Remotes are also vital in managing multiple upstream repositories, enabling developers to work across forks or contribute to external projects. For example, you might add multiple remotes—one for your team's repository and another for a public fork of the same project. By effectively naming and organizing these remotes, you can streamline workflows and maintain clarity when collaborating on complex codebases. Git remotes, when understood and utilized properly, are indispensable tools for managing distributed development and ensuring seamless collaboration.

Conclusion

Mastering Git's foundational concepts and workflows is essential for any developer or team striving for effective collaboration and robust version control. From understanding the basic principles to leveraging advanced features like aliases, tagging, and remotes, Git provides a robust framework for managing code changes efficiently. As developers interact with Git's tools and commands, they gain a deeper appreciation for its ability to streamline workflows, ensure consistency, and enhance collaboration in even the most complex projects.

By integrating Git into daily development practices, individuals and teams can unlock new levels of productivity and flexibility. The principles explored in this chapter serve as a springboard for beginners and experienced users to explore Git's full potential. Whether used for personal projects or large-scale collaborative endeavors, Git remains a cornerstone of modern software development, empowering developers to confidently build, share, and innovate.

CHAPTER 8

Advanced Git Topics

As development projects become complex, mastering advanced Git topics becomes essential for maintaining an efficient workflow. While the fundamentals of Git provide a solid foundation, diving into advanced concepts like branching, merging, and troubleshooting equips developers with the skills to handle intricate codebases and collaborative challenges. These advanced techniques allow for greater flexibility, enabling teams to work on multiple features simultaneously, resolve conflicts effectively, and maintain a clean and organized repository structure.

This chapter explores Git's advanced capabilities, focusing on branching and merging strategies, best practices to optimize workflows, and tips for troubleshooting common issues. By delving deeper into these topics, you'll learn to tackle sophisticated development scenarios confidently. Whether managing a solo project or collaborating on a large-scale enterprise application, understanding and applying these advanced Git techniques can significantly enhance your efficiency and effectiveness as a developer.

Branching

Branching is one of Git's most powerful features. It allows developers to work on different project parts simultaneously without interfering with the main codebase. A branch is a diverging path from the main commit timeline, enabling the isolation of changes. This isolation is advantageous when developing new features, fixing bugs, or experimenting with ideas, as it prevents unfinished work from disrupting the primary project.

A key benefit of branching is the ability to maintain a clean and stable main branch, often referred to as the main or master branch. Developers can create feature branches for new functionalities, hotfix branches for urgent bug fixes, or even experimental branches to test new approaches. Each branch operates independently, and changes made in one branch do not affect others unless explicitly merged. This separation ensures that the main branch remains deployable at all times.

R. L. Steelman, Jr., *Mastering Snowflake DataOps with DataOps.live*, https://doi.org/10.1007/979-8-8688-1754-0_8

The process of creating a branch is simple and efficient in Git. Using the git branch *<branch_name>* command, you can create a new branch without altering your working directory. You can switch between branches seamlessly with the *git checkout <branch_name>* or the more modern *git switch <branch_name>* command. This lightweight nature of branches in Git—unlike in older version control systems—encourages developers to use them liberally, fostering parallel development and collaboration.

Git branching is also critical for implementing specific workflows, such as Git Flow or GitHub Flow, which depend heavily on the strategic use of branches. For example, Git Flow uses separate branches for development (develop), production (main), and features (feature/*). This structured approach ensures that changes progress systematically through development and testing stages before reaching production.

One best practice when working with branches is to give them meaningful and descriptive names. For instance, instead of naming a branch feature1, use something like feature/login-authentication to communicate its purpose clearly. This clarity is invaluable in collaborative projects, as it helps team members understand the context of each branch at a glance.

Another essential consideration is regular synchronization between branches. When multiple developers work on separate branches, the codebases can diverge significantly, leading to conflicts during merging. Periodically merging changes from the main branch into your feature branch helps minimize these conflicts. Git commands like git fetch and git rebase can also help keep branches up-to-date with the latest changes from other developers.

Developers should also leverage the concept of short-lived branches. By keeping branches focused and quickly merging them back into the main branch after their purpose is fulfilled, teams can avoid the complexity and overhead of managing long-lived branches. This approach promotes an iterative development process where small, incremental changes are regularly integrated and tested.

Protected branches are essential for maintaining code integrity when working with large teams. Git allows you to restrict access to specific branches, ensuring that changes can only be made through pull requests or after meeting specific conditions, such as passing automated tests. This control helps enforce quality standards and reduces the risk of introducing errors into critical branches like main.

Git's flexibility with branches is further enhanced by tools such as visual branch viewers and GUIs, which provide a graphical representation of branch relationships. These tools are invaluable in understanding complex branching scenarios or resolving merge conflicts. By visualizing the branching structure, developers can make more informed decisions about merging and rebasing.

Branching in Git is a cornerstone of modern version control practices, enabling parallel development, efficient collaboration, and clean project management. Understanding how to effectively create, manage, and merge branches is essential for leveraging Git's full potential and fostering a productive development environment.

Merging

Merging in Git is the process of integrating changes from one branch into another. It is a fundamental operation that allows developers to consolidate work in isolated branches, ensuring that the progress made in different project areas comes together seamlessly. Whether merging feature branches into the main branch or incorporating bug fixes from a hotfix branch, the merging process ensures that contributions from various team members are integrated into a unified codebase.

The simplest type of merge is a fast-forward merge, which occurs when the merged branch has not diverged from the target branch. In this case, Git simply moves the pointer of the target branch forward to the most recent commit of the merged branch. This is efficient and clean but only possible if no other changes have been made to the target branch since the branch to be merged was created.

For more complex scenarios where the branches have diverged, Git performs a three-way merge. In this process, Git considers the latest commit on both branches and their most recent common ancestor to generate a merge commit. This merge commit reconciles the changes from both branches into a cohesive update to the target branch. While this approach retains the history of both branches, it may also lead to merge conflicts if the same parts of a file have been modified in conflicting ways.

Merge conflicts are a natural part of collaborative development and occur when Git cannot automatically reconcile differences between branches. For instance, if two developers modify the same line of code differently, Git requires manual intervention to decide which changes should be kept. Resolving conflicts involves editing the affected files to reconcile the differences and completing the merge with a commit. While conflicts can be time-consuming, they are an opportunity to review changes and ensure they align with the project's goals.

It's essential to approach merging strategically to avoid potential pitfalls. One best practice is frequently merging changes from the main branch into feature branches during development. This practice ensures that feature branches remain up-to-date with the latest changes, reducing the risk of conflicts when merging them back into the main branch. Additionally, using descriptive commit messages for merge commits helps provide context for the integrated changes.

Another consideration is the choice between merging and rebasing. While merging preserves the history of both branches, rebasing rewrites the commit history to create a linear progression of changes. Each method has benefits and trade-offs; the choice often depends on team workflows and preferences. For example, teams prioritizing a clean, linear history might favor rebasing, while those valuing a complete and accurate record of all development activity might prefer merging.

In collaborative projects, using pull or merge requests can add an extra layer of quality control to the merging process. These requests allow team members to review changes before merging into the main branch, ensuring that the new code adheres to project standards and does not introduce bugs. This process promotes accountability and fosters a culture of shared ownership and continuous improvement within the team.

Merging is a critical operation in Git that requires careful planning and execution to ensure the integrity of the codebase. By understanding the nuances of different merge strategies, resolving conflicts effectively, and adhering to best practices, teams can maximize Git's powerful capabilities to manage collaborative development efficiently.

Best Practices

Git best practices help ensure smooth collaboration, maintainable code, and an efficient development workflow. Following these practices minimizes the risk of merge conflicts, simplifies debugging, and keeps the repository organized. Whether working on a solo project or contributing to a large team, adhering to best practices enhances productivity and prevents common pitfalls.

One of the most important best practices is to commit frequently with meaningful commit messages. Frequent commits allow developers to track changes at a granular level, making it easier to identify when and where a bug was introduced. Each commit message should be clear and concise, describing the purpose of the change. A suitable format to follow is a summary in the first line, followed by a more detailed explanation if necessary. This makes it easier to understand the commit history when reviewing past changes.

Another key practice is effectively using branches. Developers should create separate branches for each feature, bug fix, or experiment instead of working directly on the main branch. This ensures the main branch remains stable and production-ready while allowing parallel development on multiple features. It is also beneficial to follow a branch naming convention, such as ***feature/add-login, bugfix/fix-header,*** or ***hotfix /security-patch***, so team members can quickly identify the purpose of a branch.

Keeping the repository clean and organized is essential for long-term maintainability. Large repositories can become cluttered with unused branches, making it challenging to manage the project efficiently. Developers should regularly delete merged or obsolete branches to keep the repository tidy. Additionally, ignoring unnecessary files by using a .gitignore file prevents unwanted files, such as build artifacts or environment-specific configurations, from being tracked.

Merging strategies also play a crucial role in practical Git usage. Developers should frequently merge changes from the main branch into their feature branches to stay updated with the latest developments and reduce merge conflicts. Additionally, when merging back into the main branch, using pull or merge requests ensures that other team members review changes before integration. Code reviews help maintain high-quality code and catch potential issues before they become problems.

Rebasing can be a powerful alternative to merging when working with feature branches. Instead of creating merge commits, rebasing rewrites the commit history to apply changes on top of the latest main branch updates. This keeps the commit history clean and linear, making it easier to follow. However, developers should avoid rebasing public branches, as it rewrites history and can cause issues for others who have already pulled the original commits.

Another best practice is to avoid committing large binary files or generated content to the repository. Git is optimized for text-based files, and large binaries can bloat the repository, slowing down operations. If binary files need to be stored, using Git LFS (Large File Storage) is recommended. Additionally, keeping commits small and focused makes reviewing changes easier and reduces the risk of introducing unintended errors.

Security should also be a consideration when using Git. Developers should never commit sensitive information like API keys, passwords, or database credentials. Instead, these should be stored in environment variables or secret management tools. If a secret is accidentally added to a commit, it should be removed immediately, and the credentials should be rotated to prevent unauthorized access.

By following Git best practices, teams can maintain a structured and efficient workflow that minimizes errors and maximizes collaboration. These practices create a development environment where changes are easy to track, issues are quickly resolved, and the repository remains organized over time.

Troubleshooting and Tips

Troubleshooting Git issues is an essential skill for developers, as problems can arise from incorrect commands, merge conflicts, or repository misconfigurations. Understanding common issues and their solutions can save time and prevent frustration. Many problems in Git can be diagnosed by carefully reading error messages and using built-in Git commands to inspect the repository's state.

One of the most common issues developers face is merge conflicts. Merge conflicts occur when two branches modify the same part of a file, and Git cannot automatically determine which version to keep. When this happens, Git marks the conflicting sections in the affected files, requiring manual intervention. The best way to resolve a merge conflict is to open the file in a text editor, review the conflicting changes, and decide which version (or a combination of both) should be kept. After resolving the conflict, the file should be staged and committed to complete the merge.

Another frequent problem is accidentally committing sensitive information, such as passwords or API keys. Suppose sensitive data is pushed to a public repository. In that case, it should be removed immediately using git filter-branch or the more efficient *git rebase* and *git push --force* to rewrite history. However, once a secret has been exposed, it should be considered compromised, and new credentials should be issued to prevent security risks.

Sometimes, developers may find themselves working in the wrong branch and committing changes unintentionally. If the changes haven't been pushed, they can be moved to the correct branch using git stash, switching to the desired branch, and then applying the stashed changes. Alternatively, if the commit should be moved, the git cherry-pick command can be used to apply a specific commit to another branch.

Another common mistake is accidentally overwriting local changes with a git pull. If local modifications were lost due to an unintended pull, Git provides recovery options. The git reflog command can be used to track previous states of the repository, allowing developers to restore lost commits or changes. Additionally, if uncommitted changes were overwritten, *git fsck --lost-found* might help recover orphaned objects.

Developers may encounter authentication or permission errors when working with remote repositories, especially when using SSH keys or personal access tokens. If Git prompts for credentials unexpectedly, checking the SSH key setup with ssh -T git@ github.com (for GitHub) or verifying the credential helper settings can resolve the issue. If using HTTPS authentication, ensuring that the correct credentials or access tokens are stored in the Git configuration can prevent authentication failures.

Another frustrating issue is when Git refuses to push due to a non-fast-forward error. This happens when the local branch is behind the remote branch, meaning someone else has pushed changes to the same branch. The safest way to resolve this is by pulling the latest changes using git pull --rebase to integrate the remote updates before pushing again. If conflicts arise during rebase, they must be resolved manually before continuing.

To prevent common Git issues, developers should adopt good habits, such as frequently committing changes, regularly pulling updates from the main branch, and reviewing changes before pushing. Using git status and git diff before committing can help ensure that only the intended changes are staged. Additionally, leveraging Git aliases can simplify complex commands, making the workflow more efficient.

Conclusion

By understanding these standard Git troubleshooting techniques and best practices, developers can work more confidently and efficiently with Git. Problems will inevitably arise, but having a solid grasp of Git's tools and commands makes resolving them much more manageable. With consistent use and experience, troubleshooting Git issues becomes second nature, allowing teams to collaborate smoothly and maintain a stable codebase.

CHAPTER 9

Introduction to DBT

The modern data landscape is evolving rapidly, with organizations generating and collecting more data than ever. However, raw data alone is not enough; it needs to be transformed into meaningful insights that drive decision-making. This is where DBT (Data Build Tool) comes in. DBT is a powerful, open-source tool designed to simplify and standardize the transformation layer of data pipelines. By enabling analysts and data engineers to write modular, version-controlled SQL-based transformations, DBT helps bridge the gap between raw data and actionable analytics.

This chapter will explore the fundamentals of DBT and its role in modern data workflows. We'll cover its core concepts, how it differs from traditional ETL tools, and why it has become a crucial component of the contemporary data stack. Whether you're an analyst looking to take control of data transformations or an engineer seeking a scalable solution for managing SQL-based models, understanding DBT will give you the tools needed to build efficient, reliable, and maintainable data pipelines.

What Is DBT?

DBT, or Data Build Tool, is an open-source analytics engineering tool enabling teams to transform raw data in their data warehouse using SQL. It bridges the gap between data engineering and analytics by allowing analysts and engineers to collaboratively build, test collaboratively, and document data models. Unlike traditional ETL (Extract, Transform, Load) tools, DBT follows an ELT (Extract, Load, Transform) paradigm, meaning it focuses solely on the transformation layer after data has already been ingested into a warehouse like Snowflake, BigQuery, or Redshift. This approach leverages the power and scalability of modern cloud data warehouses to perform transformations efficiently.

R. L. Steelman, Jr., *Mastering Snowflake DataOps with DataOps.live*, https://doi.org/10.1007/979-8-8688-1754-0_9

One of DBT's core strengths is its ability to simplify and standardize data transformation workflows. By using DBT, teams can create modular, reusable SQL-based models that are easy to understand and maintain. Instead of writing complex transformation scripts in Python or other languages, DBT allows data teams to define transformations in SQL and organize them in a structured manner. This modularity helps ensure that transformations are transparent, auditable, and scalable, reducing the risk of errors and inconsistencies in data pipelines.

Another key aspect of DBT is its focus on the analytics engineering discipline that merges traditional data engineering and analytics. Historically, data analysts often depended on engineers to build and maintain complex ETL pipelines. With DBT, analysts can take ownership of the transformation process, write SQL-based models, and follow software engineering best practices like version control, testing, and documentation. This shift democratizes data transformation and fosters a more agile, self-service approach to analytics.

DBT also enforces best practices through its robust framework. It encourages teams to follow a declarative approach, where transformations are defined as models that build on one another in a dependency graph. This makes it easy to track relationships between different datasets and ensure that transformations are executed in the correct order. Jinja, a templating language, further extends DBT's capabilities by allowing users to create dynamic SQL queries and macros that promote code reuse and consistency.

One of DBT's most valuable features is its built-in testing capabilities. Data quality is a crucial aspect of any analytics workflow, and DBT enables teams to define tests that automatically validate data integrity. These tests can check for issues like uniqueness, null values, or referential integrity, ensuring that datasets meet predefined expectations before they are used in reporting or analytics. This proactive approach helps reduce data errors and ensures stakeholders can trust the insights derived from transformed data.

Beyond testing, DBT also includes automated documentation generation. By leveraging the same models used for transformation, DBT creates clear, user-friendly documentation that provides visibility into the structure and dependencies of the data pipeline. Analysts and engineers can easily understand how data is processed, where it comes from, and how different models interact, reducing knowledge gaps and enhancing team collaboration.

DBT is highly extensible and integrates seamlessly with modern data tools and workflows. It can be used with version control systems like Git, enabling teams to track changes, collaborate on transformations, and enforce development best practices.

Additionally, DBT Cloud, the managed version of DBT, provides additional features such as a web-based UI, job scheduling, and enhanced security, making it easier for organizations to adopt DBT at scale.

As organizations continue to adopt a modern data stack, DBT has become a cornerstone technology for data transformation. Its ability to empower data teams, enforce best practices, and streamline data workflows has made it an essential tool for businesses looking to efficiently derive value from their data. Whether used in a small analytics team or as part of an enterprise-wide data platform, DBT provides the flexibility, scalability, and governance needed to drive practical analytics engineering.

In summary, DBT revolutionizes how data teams approach transformations by providing a structured, SQL-based framework that aligns with software engineering principles. Its focus on modularity, testing, and documentation enhances collaboration and trust in data processes, making it a powerful tool for modern data-driven organizations. By leveraging DBT, teams can ensure that their data pipelines are not only efficient and scalable but also maintainable and reliable in the long run.

Key Concepts in DBT

DBT (Data Build Tool) operates on core concepts, making it an effective tool for data transformation and analytics engineering. At its core, DBT enables users to define, document, and test SQL-based transformations in a structured and scalable manner. Unlike traditional ETL tools, which focus on extracting and loading data before transformation, DBT follows an ELT (Extract, Load, Transform) paradigm, emphasizing in-warehouse transformations. This means raw data is first loaded into a data warehouse, and DBT handles the transformation layer, making analytics workflows more efficient and scalable.

One of the foundational concepts of DBT is the models. Models are SQL files that define how data should be transformed within the warehouse. Each model represents a transformation step, which can be as simple as cleaning raw data or as complex as creating aggregated views for reporting. Models in DBT are highly modular, enabling users to build transformation logic step by step and maintain transparency in data processing. By structuring models effectively, teams can create reusable transformations that improve maintainability and collaboration.

Another key concept is sources and staging. In DBT, sources represent raw data tables in the warehouse, while staging models help clean and structure this data before further transformations. This layered approach ensures that transformations are not applied directly to raw data but follow a logical progression, making debugging and validation easier. Staging models often standardize column names, enforce data consistency, and use lightweight transformations before more profound transformations occur.

DBT also emphasizes testing as a fundamental practice. Unlike traditional data workflows where quality checks are often ad hoc, DBT allows users to define tests directly within their transformations. Built-in tests can check for null values, uniqueness, referential integrity, and other typical data quality issues. Custom tests can also be written using SQL, ensuring potential data anomalies are caught early in the pipeline. This focus on testing improves trust in data and prevents downstream issues.

Documentation is another core pillar of DBT. With DBT, users can directly define and store metadata about models, columns, and business logic within their project. This documentation can be automatically compiled into a web-based interface, making it easy for analysts, engineers, and stakeholders to understand how data is processed. Well-documented transformations not only improve collaboration but also ensure that institutional knowledge is preserved over time.

A unique feature of DBT is the ref() function, which enables dependency management between models. Instead of hardcoding table references, DBT allows models to refer to each other dynamically. This means when changes are made to a model, DBT can determine dependencies and execute transformations in the correct order. This feature helps build modular and scalable pipelines, as transformations are not isolated but interlinked in a structured way.

Another essential concept is incremental models, which help optimize performance when dealing with large datasets. Instead of rebuilding entire tables every time a transformation runs, DBT can process only new or updated records, significantly reducing compute costs and improving efficiency. Incremental models are particularly useful for real-time or frequently updated data, as they enable transformations to scale effectively without overwhelming the data warehouse.

Finally, DBT supports environments and version control through Git integration. Because DBT projects are structured as code, teams can apply software engineering best practices, such as code reviews, branching strategies, and CI/CD pipelines. This ensures that changes to transformation logic are tracked, tested, and deployed systematically, reducing the risk of errors.

By understanding these key concepts, data teams can fully leverage DBT's capabilities to build robust, scalable, and maintainable data transformation pipelines. With its modular approach, testing framework, and documentation tools, DBT is a game-changer for modern data engineering, enabling teams to deliver high-quality, reliable insights efficiently.

DBT Workflow and How It Fits into Data Pipelines

A well-structured data pipeline ensures reliable, scalable, and efficient data transformation, and DBT plays a key role in modern ELT workflows. Unlike traditional ETL (Extract, Transform, Load) processes, where transformations happen before loading data into the warehouse, DBT follows the ELT (Extract, Load, Transform) paradigm. This means raw data is first loaded into a data warehouse, and DBT handles the transformation within it. This approach leverages the power of cloud-based databases like Snowflake, BigQuery, and Redshift, allowing transformations to scale efficiently without moving large datasets between systems.

The DBT workflow is designed to integrate seamlessly into a modern data stack. It typically begins by extracting data from various sources, such as APIs, databases, or third-party applications, and loading it into a cloud data warehouse using tools like Fivetran, Airbyte, or Stitch. Once the raw data is available in the warehouse, DBT takes over by transforming it into a structured format suitable for analytics and reporting. The entire workflow is managed using SQL-based models, making it accessible to data analysts and engineers alike.

A core component of the DBT workflow is the development environment, where users define and refine their transformation logic. Since DBT projects are built as code, transformations are written in SQL within version-controlled repositories like Git. This allows for tracking changes, collaboration, and rollbacks when necessary. Developers work in isolated environments, making modifications without affecting production data, ensuring a smooth and controlled transformation process.

Once models are written, the next step is running DBT commands to execute transformations. The dbt run command processes the models, applying transformations in the correct order based on dependencies defined using the ref() function. This ensures upstream models are executed before dependent models, preventing errors and inconsistencies. DBT also provides commands like dbt debug to validate connections, dbt test to run data quality checks, and dbt snapshot to track historical changes in slowly changing dimensions.

Testing is an integral part of the DBT workflow. With dbt test, users can validate their data transformations by applying predefined or custom tests. Standard tests include checking for null values, uniqueness constraints, and referential integrity. These automated tests help catch data issues early, ensuring high-quality outputs before they reach business users. Unlike traditional pipelines that rely on manual QA, DBT enables a test-driven development approach for data, making analytics engineering more robust.

Once transformations are validated, DBT supports documenting models to improve collaboration and transparency. Using YAML files, users can describe models, columns, and relationships, making it easier for analysts and engineers to understand how data flows through the pipeline. DBT generates a web-based documentation interface (dbt docs generate), allowing teams to visually explore data lineage and dependencies. This built-in documentation feature ensures institutional knowledge is preserved and accessible to all stakeholders.

Another critical aspect of the DBT workflow is orchestration and automation. While DBT runs transformations, it is typically integrated with workflow orchestration tools like Apache Airflow, Dagster, or Prefect to schedule and automate pipeline execution. By defining dependencies and execution schedules, teams can ensure that data transformations occur regularly, keeping analytics dashboards current. Automated orchestration also enables monitoring and alerting, helping teams quickly detect and resolve pipeline failures.

DBT also plays a crucial role in data lineage tracking. Because every transformation step is defined within DBT models, it becomes easy to trace how a particular dataset was derived. This transparency is vital for debugging issues, ensuring compliance, and understanding the impact of changes before deploying them to production. Data lineage tracking is beneficial in complex analytics workflows, where multiple transformations contribute to business-critical insights.

DBT's modular approach is a game-changer for teams managing large-scale data pipelines. By breaking down transformations into reusable and interdependent models, teams can improve maintainability, reduce redundancy, and scale their analytics infrastructure efficiently. Instead of writing long, monolithic SQL scripts, DBT encourages a structured approach where transformations are built incrementally, promoting a more organized and scalable workflow.

Ultimately, DBT enhances the agility of data teams by aligning data transformation workflows with software engineering best practices. With its SQL-based transformations, version control, testing, documentation, and orchestration capabilities, DBT streamlines the entire data pipeline, making it easier to build, maintain, and scale analytics solutions. By integrating DBT into modern ELT pipelines, organizations can deliver faster, more reliable insights while ensuring data quality and governance across the board.

Setting Up DBT

Setting up DBT requires a combination of environment configuration, installation, and connection to a data warehouse. The process varies slightly depending on whether you're using DBT Cloud or DBT Core, but the fundamental steps remain the same. DBT Cloud is a managed solution that simplifies setup and execution, while DBT Core is the open-source command-line version that provides more flexibility. Choosing the correct version depends on the team's needs—DBT Cloud is great for collaboration and ease of use, whereas DBT Core offers greater control for engineering-heavy workflows.

The first step in setting up DBT is installing DBT Core if you choose the open-source option. DBT is written in Python, requiring Python and pip (Python's package manager) to be installed. Running the command pip install dbt-core installs the core DBT package, but additional adapters are required based on the data warehouse being used. For example, to connect DBT to Snowflake, the command pip install dbt-snowflake is needed. Similarly, adapters exist for BigQuery, Redshift, and other platforms.

Once DBT is installed, the next step is to initialize a DBT project. This is done using the command dbt init project_name, which generates a project directory with essential folders and configuration files. The key components include a models/ folder for SQL transformations, a profiles.yml file for connection settings, and a dbt_project.yml file for project-level configurations. These files define DBT's interaction with the data warehouse and structure transformation logic.

A crucial part of the setup process is configuring the connection to a data warehouse. DBT stores authentication credentials and connection settings in the profiles.yml file, typically located in the ~/.dbt/ directory. Users must provide details such as the database type, hostname, user credentials, and warehouse settings. Some warehouses, like Snowflake, support authentication via OAuth or key-pair authentication for added security. Running dbt debug helps verify the connection and troubleshoot any authentication issues.

After establishing a connection, testing the setup by running a basic DBT command is essential. The dbt debug command checks the environment and confirms that DBT can connect to the warehouse. A successful output indicates that DBT is ready to execute transformations. Additionally, running dbt run on an initial model confirms that transformations execute correctly. If any errors occur, they usually stem from misconfigured credentials or incorrect database permissions.

Once the basic setup is verified, teams can organize their DBT project structure based on best practices. Within the models/ directory, it's common to create subdirectories like staging/, intermediate/, and marts/ to separate different layers of transformations. Staging models contain raw data, intermediate models apply business logic, and marts contain final tables for reporting. This structured approach improves maintainability and ensures a clear transformation workflow.

~/.dbt/profiles.yml

```yaml
my-snowflake-db:
  target: dev
  outputs:
    dev:
      type: snowflake
      account: [account id]

      # User/password auth
      user: [username]
      password: [password]

      role: [user role]
      database: [database name]
      warehouse: [warehouse name]
      schema: [dbt schema]
      threads: [1 or more]
      client_session_keep_alive: False
      query_tag: [anything]

      # optional
      connect_retries: 0 # default 0
      connect_timeout: 10 # default: 10
      retry_on_database_errors: False # default: false
      retry_all: False  # default: false
      reuse_connections: True # default: True if client_session_keep_alive is False, otherwise None
```

Figure 9-1. *Sample profiles.yml for Snowflake*

To enhance collaboration, DBT projects are typically integrated with Git for version control. Running git init inside the DBT project directory initializes a repository, allowing teams to track changes, create branches, and collaborate efficiently. DBT Cloud provides built-in Git integration, making it easier for teams to manage versioning and deployments. Using Git ensures that all transformations are documented, making it easier to roll back changes if needed.

The setup process is more streamlined for teams using DBT Cloud. After signing up for a DBT Cloud account, users can create a new project and configure the connection through a guided interface (Figure 9-2). DBT Cloud handles authentication, scheduling, and execution, reducing the need for manual configuration. Additionally, it provides a web-based development environment, removing the need to install DBT locally. This makes it a preferred choice for teams that want a managed solution with minimal setup overhead.

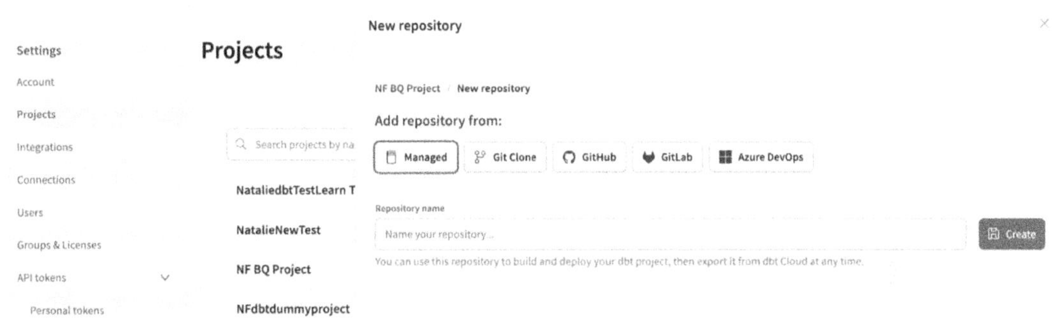

Figure 9-2. *New DBT Cloud Project*

Finally, after setting up DBT, it's essential to establish CI/CD (Continuous Integration and Continuous Deployment) workflows for automated testing and deployment. Many teams integrate DBT with orchestration tools like Airflow or Prefect to schedule runs and monitor pipeline performance. Setting up automated testing with dbt test ensures that transformations meet data quality standards before they reach production. With a well-structured setup, DBT enables scalable, reliable, and maintainable data transformations that fit seamlessly into modern data pipelines.

Developing and Running DBT Models

Developing and running DBT models is at the core of leveraging DBT for data transformation. A DBT model is an SQL file that defines how raw data should be transformed into usable, structured outputs. These models enable analysts and data engineers to write modular, version-controlled transformations that can be easily tested, documented, and executed. DBT provides a structured way to build scalable and maintainable data pipelines by defining models using SQL and YAML configurations.

A DBT model typically lives in the DBT project's models/directory and follows a standard hierarchy. Models can be categorized into staging models, intermediate models, and marts (final reporting tables). Staging models clean and standardize raw data, intermediate models apply business logic, and marts contain aggregated or final datasets ready for reporting and analytics. Structuring models ensures that transformations follow a logical and maintainable process.

To create a DBT model, users simply write an SQL query inside a .sql file. For example, a staging model might clean up customer data by standardizing names and removing duplicates. An intermediate model might join multiple tables to create a unified customer profile, and a final mart model could generate aggregated revenue metrics per customer. DBT allows users to reference other models within SQL using the {{ ref('model_name') }} function, which ensures that dependencies are handled automatically.

Once models are defined, they need to be compiled and executed. The command dbt run compiles all models into executable SQL queries and runs them against the data warehouse. DBT translates the defined transformations into optimized SQL queries that execute efficiently in platforms like Snowflake, BigQuery, or Redshift. Running models in this way ensures that transformations are applied consistently and reproducibly across environments.

DBT also supports incremental models, which optimize performance by processing only new or changed data rather than recomputing entire tables. Incremental models use the is_incremental() function to check whether a model is being run for the first time or as an update. This approach significantly improves efficiency, especially for large datasets, as only new records are processed instead of reloading the entire table.

Testing is a crucial part of DBT model development. The command dbt test runs predefined tests to ensure that data meets expectations. Users can define custom tests in YAML files to check for null values, unique constraints, referential integrity, and other conditions. Ensuring data quality through testing helps prevent downstream issues in analytics and reporting.

Another key feature of DBT is model documentation. The command dbt docs generates a dynamic documentation site that describes all models, their dependencies, and the transformations applied. Users can add descriptions and metadata to their models in YAML files to improve readability and maintainability. The ability to generate interactive documentation ensures stakeholders understand how data is being transformed.

DBT also supports model configurations, which allow users to define settings such as materialization type (view, table, or incremental), clustering, and partitioning strategies. These configurations help optimize query performance and storage efficiency. Configurations can be set within models or globally in the dbt_project.yml file, making enforcing best practices across an organization easy.

For large-scale projects, modularization of DBT models is essential. Breaking transformations into smaller, reusable models improves maintainability and reduces complexity. By keeping logic separate, teams can debug and update transformations without affecting downstream processes. This modular approach also aligns well with software engineering principles, making DBT a robust tool for data transformation.

After developing and running DBT models, the final step is integrating them into orchestration workflows. Many teams use tools like Apache Airflow, Prefect, or Dagster to schedule DBT runs and monitor execution. By automating model execution and incorporating CI/CD pipelines, teams can ensure that transformations are applied consistently and reliably in production environments. This makes DBT a powerful component of the modern data stack, bridging the gap between raw data ingestion and actionable insights.

Testing and Documentation in DBT

Testing and documentation are two fundamental aspects of DBT that ensure data integrity and maintainability in modern data pipelines. Testing helps detect issues in transformations before they reach production, while documentation improves transparency and collaboration among data teams. By incorporating both into a DBT workflow, organizations can create reliable, well-documented data models that foster trust and usability.

Testing in DBT

DBT provides built-in testing capabilities that allow users to define and enforce data quality standards. Tests in DBT are SQL-based assertions that verify whether data meets expected conditions. These tests can be applied to individual columns or tables to catch issues such as missing values, duplicate records, or referential integrity problems before impacting downstream analytics.

DBT supports two primary types of tests: generic tests and singular tests. Generic tests are predefined checks, such as uniqueness, not-null constraints, accepted values, and referential integrity. These tests are declared in YAML files and automatically applied to models when executing dbt test. Singular tests, on the other hand, are custom SQL queries written by users to validate specific business rules or data conditions.

For example, a generic test can be added to ensure that a customer_id column is unique and never null:

Listing 9-1. YAML: Generic dbt test

```
version: 2
models:
  - name: customers
    columns:
      - name: customer_id
        tests:
          - unique
          - not_null
```

Singular tests allow for more flexibility in defining business logic. A user can create a custom test file in the tests/ directory, such as high_value_customers.sql, to check that all high-value customers have a valid email:

Listing 9-2. SQL: In-line custom SQL test

```
SELECT *
FROM {{ ref('customers') }}
WHERE total_spent > 10000 AND email IS NULL
```

Running the dbt test executes all defined tests, flagging any data quality issues. The ability to apply tests directly within the transformation layer ensures that problems are caught early, reducing the risk of inaccurate reporting and analysis.

Documentation in DBT

DBT also excels at generating documentation that helps teams understand their data models. Documentation clearly explains what each model does, its dependencies, and the transformations applied. By keeping documentation up-to-date and accessible, teams can ensure consistency in how data is used across an organization.

Documentation in DBT is stored as YAML metadata within the project. Users can add descriptions to models, columns, and tests, making it easier for others to understand the purpose and logic behind each transformation. A simple example of adding documentation to a DBT model looks like this:

Listing 9-3. DBT documentation example

```
version: 2
models:
  - name: orders
    description: "This table contains order data from the e-commerce
    platform."
    columns:
      - name: order_id
        description: "Primary key for the orders table."
        tests:
          - unique
          - not_null
      - name: order_date
        description: "Timestamp of when the order was placed."
      - name: total_amount
        description: "Total value of the order, including taxes and
        discounts."
```

Once documentation is defined, running dbt docs generates and compiles it into an interactive website that visualizes model dependencies, descriptions, and test results. This documentation can be hosted internally and shared across teams for better visibility into the data pipeline.

Testing and Documentation Together

Combining testing and documentation in DBT ensures that transformations are both reliable and well-documented. Testing helps maintain data accuracy, while documentation enables data teams to understand how models are structured and why certain transformations are applied. Together, these features promote data quality, governance, and collaboration.

A best practice is to enforce automated testing in CI/CD workflows. By integrating dbt test into deployment pipelines, teams can automatically validate data quality before pushing changes to production. This prevents insufficient data from propagating and reduces manual intervention in troubleshooting.

Similarly, keeping documentation up-to-date is crucial. Teams should treat documentation as a living artifact, updating descriptions whenever models change. Regularly reviewing documentation ensures that new team members can onboard quickly and that stakeholders trust the data.

By leveraging DBT's robust testing and documentation features, organizations can create a well-governed data transformation process that ensures both quality and transparency. These capabilities make DBT an indispensable tool for modern data teams, enabling them to build scalable, reliable, and well-documented analytics pipelines.

Version Control and Collaboration with DBT

Version control is essential for managing changes to code, tracking the history of modifications, and enabling seamless collaboration within teams. In the context of DBT, version control helps data teams collaborate efficiently on complex data transformation projects while ensuring the integrity and consistency of the work. By integrating DBT with version control systems like Git, data teams can work together without the risk of overwriting each other's changes or losing historical work.

In DBT, version control typically involves managing both the SQL-based transformation models and the configuration files that define the project's structure. Using Git, developers can create branches for new features or bug fixes, track changes over time, and review code through pull requests. This enables a collaborative approach where team members can contribute, review each other's work, and discuss improvements or modifications before integrating them into the main codebase. GitHub,

GitLab, and Bitbucket are commonly used for this purpose, offering platforms to host repositories and streamline workflows with features such as pull requests, issue tracking, and code reviews.

The use of branches and pull requests allows teams to isolate features or bug fixes and work on them independently. For example, one team member may develop a new model to track customer behavior, while another focuses on optimizing an existing transformation for performance. Both can work in parallel without impacting each other's progress. Once the work is complete, changes can be reviewed and merged into the main branch, ensuring that only stable and tested code is pushed to production.

Moreover, version control with DBT makes it easier to handle rollback and traceability in case of issues. When a problem arises, having a detailed version history makes it possible to identify what change introduced the issue and revert to a previous, stable version of the model. This is particularly important in production environments, where stability and reliability are critical. Additionally, version control helps enforce best practices like keeping commit messages clear and descriptive, ensuring that the history of changes is both traceable and understandable to anyone reviewing the project later.

Collaboration extends beyond individual team members working in parallel; it also includes cross-functional collaboration with stakeholders. By using version control and collaboration tools, data teams can include non-technical stakeholders, such as data analysts or business users, in the process. These stakeholders can track progress, raise concerns, or even suggest changes. By encouraging collaboration and involving diverse perspectives, teams can build better data pipelines that are aligned with business needs and deliver valuable insights.

One advantage of version control in DBT is that it supports collaboration in a way that reduces friction between data teams and IT operations. For example, DBT's integration with CI/CD tools and version control systems means that when a model is updated, it can be automatically tested, deployed, and monitored. This tight integration between code, testing, and deployment allows data teams to manage their work more efficiently and ensures that updates are thoroughly vetted before being deployed to production. This minimizes risk and reduces the time spent manually testing and deploying changes.

Finally, version control with DBT fosters knowledge sharing. With all the code and documentation stored in a centralized version-controlled repository, new team members can quickly get up to speed by reviewing past changes and understanding the evolution of the project. This accessibility to historical work and project context promotes knowledge sharing within teams and makes onboarding new developers or analysts more efficient. As teams grow and projects scale, version control ensures that everyone stays aligned and that knowledge is preserved for future reference.

In summary, version control and collaboration in DBT enable data teams to work more efficiently, reduce errors, and ensure that their data transformation processes are transparent and accountable. It not only facilitates seamless collaboration but also enhances the overall reliability, stability, and traceability of the entire project. By embracing version control as part of the DBT workflow, teams can build scalable and high-quality data pipelines that are easily maintained and continuously improved over time.

Conclusion

The implementation of version control and collaboration tools within DBT transforms the way data teams manage, collaborate, and scale their workflows. By integrating version control systems like Git, teams can maintain organized, traceable, and efficient processes that reduce risks, enhance collaboration, and improve the overall quality of their data pipelines. These practices enable teams to iterate quickly while ensuring that data transformations are both stable and reproducible, which is crucial for maintaining the integrity of large-scale data projects.

Moreover, version control and collaboration go beyond just coding practices; they foster a culture of transparency and shared responsibility. Through version control, all team members can contribute, review, and track changes, ensuring that everyone stays aligned with project goals and objectives. This collaborative environment not only enhances the technical aspects of the work but also strengthens cross-functional communication and knowledge sharing, creating a more resilient and adaptable data pipeline development process.

CHAPTER 10

Advanced DBT Techniques and Best Practices

This chapter is designed to take your DBT skills to the next level, helping you optimize and scale your transformations in real-world, complex environments. While DBT's core functionalities are powerful for managing data pipelines and transformations, understanding advanced concepts and techniques will allow you to maximize its potential. This chapter explores the intricacies of building high-performance models, managing large-scale DBT projects, and integrating advanced features like custom macros, incremental loading, and testing strategies to improve the reliability and speed of your data pipeline.

This chapter enhances the efficiency and scalability of your DBT projects and emphasizes best practices that foster collaboration, maintainability, and continuous integration in a data environment. With DBT becoming central to many modern data operations, mastering these advanced techniques ensures your workflows are optimized, your teams are aligned, and your data transformation models are robust and production-ready. Whether you're managing enterprise-level deployments or fine-tuning incremental models, these best practices and techniques will empower you to handle data transformations at scale with precision and ease.

R. L. Steelman, Jr., *Mastering Snowflake DataOps with DataOps.live*, https://doi.org/10.1007/979-8-8688-1754-0_10

Optimizing DBT Models for Performance

Optimizing DBT models for performance is crucial when working with large datasets or in environments where speed is critical. One of the first steps in ensuring your DBT models run efficiently is to focus on query performance. Complex queries or transformations that operate on millions or billions of rows can lead to long execution times, which can impact your entire data pipeline. DBT's SQL-based transformations mean that optimizing SQL queries is key to improving performance. By focusing on strategies like indexing, partitioning, and optimizing joins, you can minimize the time required for transformations to run.

When working with large datasets, consider utilizing incremental models in DBT. Instead of reprocessing all the data every time a model runs, incremental models process only the new or updated records since the last successful run. This is especially beneficial for tables with large volumes of data where a full refresh could take a significant amount of time. With incremental models, DBT will only update the rows that have changed, thus reducing the overall workload and improving the performance of the transformation process. When using incremental models, always ensure that you have an efficient strategy for identifying the records that need to be processed, often through date fields or unique identifiers.

Another critical aspect of optimizing DBT models for performance is minimizing the number of transformations required. In many cases, complex data transformations involve multiple steps that are executed in a series of dependent models. While this modular approach promotes flexibility and readability, it can also create unnecessary overhead. Instead of chaining together numerous models, it may be more efficient to consolidate transformations into fewer models or steps. By doing so, you minimize the need for intermediate table creation and reduce the amount of time DBT needs to spend moving data through the pipeline.

SQL performance tuning also plays an essential role in the efficiency of DBT models. Even minor adjustments to your SQL queries can lead to significant performance gains. Using Common Table Expressions (CTEs), for example, can help break down complex transformations into smaller, more manageable chunks. However, excessive use of CTEs can sometimes lead to performance degradation, particularly in large queries. It's also essential to use aggregate functions wisely—calculating aggregations early in the process can save significant computation time later. Additionally, optimizing the logic used in joins can prevent unnecessary scans of large tables, reducing the time spent on each transformation step.

One key aspect of performance optimization in DBT is how you handle data partitioning. Partitioning tables allows you to divide a large dataset into smaller, more manageable chunks, typically based on time intervals like days, months, or years. This can make it easier to query subsets of data and improve performance, especially for incremental models. For example, if your DBT models are based on daily logs, partitioning the data by day can help your queries focus on a smaller subset of records, thus speeding up transformations. Choosing partitioning keys carefully is vital to ensure that they align with how your data is queried, allowing the system to leverage the partitioned data more effectively.

Using materializations strategically can also have a significant impact on performance. In DBT, materializations control how models are built and stored in the database. The default materialization is the table, where DBT creates a physical table for each model. However, DBT also supports other materializations, such as views, incremental models, and ephemeral models. Choosing the right materialization for each model in your project can lead to better performance. For example, using views for models that don't require frequent updates or are simple queries can save time and computational resources compared to materializing them as tables. On the other hand, incremental models are ideal for large datasets where only a subset of records is updated.

Another technique for optimizing DBT models is query profiling and benchmarking. DBT's built-in capabilities allow you to analyze the execution time of each model and track the performance of each run. By profiling queries and observing where the bottlenecks occur, you can pinpoint areas that need optimization. If specific queries are consistently slow, it might be necessary to rewrite them using more efficient SQL patterns or consider different data partitioning and indexing strategies. Regular benchmarking can help you keep track of performance improvements over time and ensure that your models continue to run efficiently as your data grows.

For highly complex transformations, consider using parallel processing to split the workload and reduce execution times. DBT can work with databases and data warehouses that support parallel query execution, allowing multiple tasks to be run concurrently. This parallelization speeds up data processing, especially when numerous independent tasks can be executed in parallel. It's important to ensure that your database is configured to fully take advantage of this capability, especially in cloud environments like Snowflake or BigQuery, which are optimized for handling parallel queries at scale.

Finally, monitoring and alerting are critical components of optimizing DBT models in the long run. Setting up alerts for failed or delayed models ensures that you can quickly respond to issues before they affect downstream processes. You can also monitor model performance over time using DBT's built-in tools or third-party monitoring tools. Establishing clear performance thresholds and regularly reviewing performance data helps you stay on top of potential issues before they become bottlenecks. Reviewing and adjusting performance optimization strategies as your data grows and new features or enhancements are added to DBT is also essential.

As part of ongoing performance optimization, it's crucial to regularly evaluate hardware and cloud resources. If you're working in a cloud data warehouse, monitor your resource usage to ensure that you have sufficient compute and storage capacity for your models. In some cases, you may need to scale your resources up or down based on demand. Understanding how your data and transformations affect resource usage can help prevent performance slowdowns, especially during peak times when large-scale data loads and transformations are happening.

Regularly refactoring DBT models is another helpful practice in the optimization process. As your project grows, it's easy for models to become inefficient due to increasing complexity or accumulated changes. Periodically refactoring your DBT models can lead to improved performance and maintainability. This includes optimizing SQL code, consolidating redundant models, and removing unnecessary transformations. Over time, this process helps ensure that your DBT project remains agile and scalable and that performance continues to meet business needs as the volume and complexity of your data increase.

By combining these strategies—incremental models, SQL optimization, partitioning, materializations, and regular monitoring—you can significantly enhance the performance of your DBT models. Performance optimization is an ongoing process that requires careful planning, continuous analysis, and iteration. Ultimately, well-optimized DBT models lead to faster data transformations, more reliable data pipelines, and better overall system performance, enabling data teams to deliver high-quality data to stakeholders in a timely manner.

Advanced Data Transformations with DBT

Advanced data transformations with DBT (Data Build Tool) offer powerful ways to manipulate, enrich, and analyze data within a modern data stack. DBT provides a flexible environment where users can write and manage complex transformations using SQL, allowing for scalable and efficient data pipelines. This section focuses on advanced techniques that can be utilized in DBT to streamline data workflows and leverage the full potential of the tool. These techniques help data teams handle complex business logic, data quality checks, and scalable processing while ensuring consistency across their data pipeline.

One of the foundational aspects of advanced data transformations is modularization. Instead of writing large, monolithic SQL queries, DBT encourages the use of smaller, reusable models that can be assembled together. This modular approach not only makes code more readable and maintainable but also facilitates debugging and testing. In DBT, each model can serve as a building block, which means complex transformations can be broken down into more straightforward, manageable steps. This approach also helps with version control and collaboration, as different team members can work on individual models independently before merging them into the final pipeline.

Another advanced technique is using CTEs (Common Table Expressions) within DBT models. CTEs allow you to structure SQL queries in a more readable and organized manner, especially when performing complex calculations or aggregations. CTEs can be used to create temporary result sets that can be referenced multiple times in the same query, reducing duplication and improving query performance. While CTEs are generally supported by most SQL engines, in DBT they can be particularly useful when working with large datasets, as they can break down complex transformations into smaller, more efficient parts, which helps optimize performance.

Window functions are another powerful feature within DBT that can be used to perform advanced transformations. Window functions allow you to perform calculations across a set of table rows related to the current row without having to group the data. For example, running totals, moving averages, and ranking operations can be easily achieved using window functions. In DBT, window functions are often used to simplify complex data manipulations that would otherwise require self-joins or subqueries. By using window functions, you can achieve the same result with more efficient and optimized SQL queries, reducing query execution time and improving overall system performance.

For advanced users, custom macros in DBT provide a way to encapsulate reusable logic that can be applied across multiple models. Macros are SQL functions that allow you to write complex logic once and reuse it across different models, improving code maintainability and consistency. For example, write a macro for a commonly used calculation, such as calculating fiscal years or handling certain types of aggregations. Macros can also be parameterized, meaning you can pass in different arguments to customize their behavior for other models or use cases. This makes DBT an even more powerful tool for data teams working on complex, dynamic datasets.

In addition to custom macros, Jinja templating is a powerful feature in DBT that allows you to inject logic into your SQL code. Jinja is a templating engine that can be used to dynamically generate SQL queries, enabling you to write more flexible and parameterized queries. With Jinja, you can conditionally execute SQL code based on parameters, making your models more dynamic and reusable. For example, you can use Jinja to build complex SQL statements that vary based on the environment, such as development, staging, or production, or based on different inputs like time intervals or specific business rules.

Another advanced transformation technique is data testing and quality checks. DBT provides a robust testing framework that allows users to ensure the accuracy and integrity of their data as it flows through the pipeline. You can write tests to check for null values, duplicate records, data type mismatches, and other data quality issues. These tests can be run during the DBT execution process, allowing data engineers to catch potential problems quickly. Advanced users can also write custom tests to handle more complex scenarios, such as ensuring data consistency across multiple tables or verifying that business rules are adhered to during transformations.

Handling slowly changing dimensions (SCDs) is an everyday use case in advanced data transformations. DBT can be configured to manage SCDs using incremental models. Slowly changing dimensions refer to attributes in data that change infrequently, such as customer details or product information. In DBT, you can implement SCD types (like Type 1 for overwriting old data or Type 2 for tracking historical changes) using incremental models. By updating only the new or changed records, incremental models reduce processing time, especially when dealing with large datasets, and make it easier to handle changes over time.

For datasets that require complex joins or multi-step transformations, DBT supports the use of temporary tables. Temporary tables allow you to store intermediate results without committing them to the final database tables. This can be useful when dealing with intricate transformations that require several stages of data processing. By using temporary tables, you can break down the transformation process into logical steps, reducing the need for complex subqueries or multi-table joins. This approach can make your queries more efficient, improving both execution time and readability.

Advanced DBT users often leverage parallel processing for larger datasets to optimize the use of hardware resources. In cloud-based data warehouses like Snowflake, BigQuery, or Redshift, DBT can take advantage of the native parallel processing capabilities to run multiple queries simultaneously. This can dramatically speed up the data transformation process, especially when working with numerous independent transformations that can run in parallel. By configuring your DBT models to take advantage of parallel processing, you ensure that your workflows are as efficient as possible, reducing the time required for data transformations and improving overall system performance.

As DBT models become more complex, managing and organizing them efficiently becomes increasingly important. DBT's directory structure allows for organizing models into logical groupings, which is critical for large-scale projects. By maintaining a clean directory structure with well-named models, tests, and macros, you can keep your project maintainable even as it grows. Advanced DBT users often create subdirectories for different stages of the data pipeline, such as raw, staging, and analytics layers, ensuring a clear separation of concerns. This level of organization makes it easier to collaborate with team members, perform updates, and track progress across different parts of the data pipeline.

Lastly, performance tuning is an essential practice for advanced DBT transformations. As your data pipeline scales, performance issues may arise, especially when working with massive datasets. In addition to optimizing SQL queries, techniques such as indexing, partitioning, and leveraging database-specific performance features can make a significant difference. DBT provides the ability to track the performance of each model and offers insights into which steps in the pipeline are taking the longest. By continuously monitoring and adjusting your transformations, you can ensure that your DBT project remains performant and scalable as data volumes increase.

In conclusion, advanced data transformations in DBT enable data teams to handle a wide range of complex use cases with ease. By leveraging modularization, window functions, CTEs, custom macros, and incremental models, DBT empowers teams to create flexible, scalable, and efficient data pipelines. Whether you're managing slowly changing dimensions, performing complex joins, or ensuring data quality, DBT's powerful features make it easier to implement best practices and optimize performance in large-scale data transformation workflows.

Customizing DBT with Macros and Jinja

Customizing DBT with macros and Jinja is one of the most powerful features that DBT offers to enhance flexibility, reusability, and efficiency in data transformation workflows. Macros in DBT allow users to encapsulate reusable logic into functions that can be applied across multiple models. These macros can simplify complex transformations by reducing redundancy and increasing maintainability. Additionally, macros can help handle logic that is specific to your data transformation pipeline, such as conditional operations, dynamic table generation, or custom business logic. This modular approach is key to writing efficient and readable code in DBT.

At the core of DBT's customization capabilities is Jinja, a templating engine that allows for dynamic SQL generation. Jinja enables users to write SQL code that adapts to different inputs or conditions, creating flexible and reusable code. Jinja expressions are included within DBT models, allowing the users to write dynamic SQL statements that can be parameterized and customized. For instance, Jinja allows you to execute a block of SQL only when certain conditions are met or loop through a list of items to create a series of SQL statements. This capability is particularly helpful when working with complex logic that would otherwise require multiple models or manual adjustments.

Macros in DBT, powered by Jinja, allow you to define and reuse SQL functions or transformations. These macros can be stored in a central location and invoked across multiple DBT models, making them ideal for handling recurring tasks like data formatting, standardizing column names, or applying consistent transformations. For example, a macro could be written to perform date formatting in a particular way, and this macro can then be used in every model that needs this transformation. This reduces code duplication and ensures consistency across the project. Macros can be invoked with arguments, which makes them flexible and adaptable for various use cases.

One of the key benefits of using Jinja in DBT is the ability to create conditional logic in your SQL models. Through Jinja templates, DBT models can change based on variables, inputs, or parameters passed during execution. For example, you can define a variable in the DBT project configuration file or pass it via the command line and use it within your models to modify the behavior of the transformations. This feature allows for a high degree of customization, such as running certain models only on specific days or excluding certain tables based on the environment (e.g., staging versus production).

Another powerful feature of Jinja within DBT is the ability to loop through lists or arrays to dynamically create queries. This is particularly useful when you need to perform similar transformations on multiple tables or datasets. For example, you could loop through a list of tables and apply the same transformation to each one, without needing to write separate models for every table. This can significantly reduce the amount of code you need to maintain and simplify complex workflows. Additionally, the ability to use loops in conjunction with DBT's macros makes it easy to create dynamic, parameterized transformations.

Macros and Jinja also allow for advanced configuration and dynamic model generation. For example, you can use Jinja templates to create SQL that targets different schema names based on the environment or user preferences. This dynamic capability can be incredibly powerful in large organizations with multiple environments (e.g., development, staging, production) or in projects where the schema varies based on customer needs or specific data sources. With macros, DBT can adapt to different configurations seamlessly, reducing the manual effort of rewriting models for different environments or datasets.

In addition to parameterization, Jinja can be used in conjunction with DBT hooks to trigger specific actions before or after running a DBT model. For instance, you can use a hook to automatically clean up temporary tables or run custom scripts before executing a model. This gives you an added layer of control over your DBT pipeline, ensuring that your transformations are not only efficient but also properly managed. By incorporating hooks into your workflow, you can ensure that all the necessary steps are taken automatically, reducing manual intervention and increasing the reliability of your data pipeline.

DBT also offers the ability to define custom schemas and tables dynamically using Jinja templates. When working with large-scale data pipelines, you may need to create a new schema or set of tables dynamically based on business requirements. This can be accomplished with Jinja by embedding template logic that modifies the SQL code generated by DBT. For example, using Jinja, you could create models that are dynamically named or organized based on specific business rules or conditions, which can help streamline model management, particularly in large data environments with a wide array of tables and schemas.

An interesting application of Jinja and macros is in data lineage tracking. With custom macros, you can define relationships between various models, ensuring that the necessary dependencies are correctly mapped. For example, if one model depends on the output of another, you can use macros to dynamically track these dependencies. This can help improve the clarity and manageability of your DBT project, as data lineage is automatically captured as part of the model development process. When using macros for data lineage, it becomes easier to visualize and understand the flow of data through your pipeline, which can be invaluable for debugging, auditing, and optimizing your transformations.

The modularity of Jinja and macros in DBT also leads to improved collaboration among teams. By encapsulating common logic in macros, teams can work more independently, without worrying about duplicating code or making conflicting changes to SQL queries. For example, one team could focus on creating a set of macros for handling specific types of transformations, while another team could work on building models using those macros. This separation of concerns helps teams maintain focus on their tasks, reduces overlap, and ensures that changes made by one team don't negatively affect the work of another.

Finally, the use of macros and Jinja templates can make DBT more scalable. As data projects grow and become more complex, the need for reusable and configurable code increases. Macros and Jinja provide a scalable solution to manage complex transformations at scale. Whether you need to handle multiple environments, large datasets, or changing business requirements, DBT's customization options allow you to adapt your workflows and models to meet new challenges. This flexibility helps ensure that your DBT project can scale as your organization's data needs evolve, making DBT a future-proof tool for modern data teams.

In summary, customizing DBT with macros and Jinja offers a powerful way to optimize your data transformation workflows. These tools allow you to create flexible, reusable, and dynamic SQL models, improving code efficiency, maintainability, and performance. By leveraging Jinja's templating engine and DBT's macro system, you can build sophisticated transformations, enhance collaboration across teams, and scale your data pipeline to meet growing demands. Whether you're simplifying complex transformations, enabling parameterization, or managing data lineage, macros and Jinja are essential tools for advanced DBT users looking to optimize their workflows.

Leveraging DBT's Advanced Testing Capabilities

Leveraging DBT's advanced testing capabilities is crucial for ensuring the integrity and accuracy of your data pipeline. DBT's built-in testing framework allows data teams to define custom tests, automate data validation, and improve the quality of their transformations. The testing framework offers several predefined tests, such as checking for null values, uniqueness, or referential integrity, which can be easily applied to models. These tests ensure that the transformed data adheres to business rules, stays consistent over time, and meets the necessary quality standards.

One of the most powerful features of DBT's testing framework is the ability to define custom tests. While DBT provides a set of standard tests, organizations often have unique requirements or complex business rules that need specific validation. In such cases, users can define custom tests using SQL queries to check data for particular conditions. For instance, a custom test could validate that a specific date field follows the correct format or that the total value in a transaction table does not exceed a certain threshold. Custom tests add an additional layer of flexibility and allow teams to define precisely what constitutes "good" data for their specific use cases.

DBT's testing capabilities extend to data relationships and referential integrity as well. For example, DBT can check that foreign key relationships between tables are valid and that no orphaned records exist. Referential integrity testing ensures that all foreign key relationships point to valid primary keys, which helps to maintain the consistency and accuracy of your data model. In a large-scale data project, where different data sources are combined into a central data warehouse, these tests are essential for identifying data mismatches and ensuring that all tables are aligned with the business logic.

Another critical aspect of DBT's testing functionality is the ability to test data across different environments. DBT allows for testing on various environments, such as development, staging, and production, enabling teams to ensure that their models behave as expected before being deployed to production. This feature helps catch issues early in the pipeline, preventing faulty data from reaching end users. It also enables teams to test their transformations with realistic data and make adjustments before pushing any changes live.

For teams that need to automate testing in their CI/CD (Continuous Integration/Continuous Deployment) pipelines, DBT provides seamless integration with popular CI/CD tools like GitHub Actions, Jenkins, and CircleCI. By automating tests during the development process, teams can quickly identify failures and make adjustments in real time. This level of automation ensures that data quality is constantly monitored and that models are validated at every stage of development. Automated testing also reduces the need for manual intervention, leading to faster deployment cycles and fewer errors in production.

Data quality testing in DBT can also include a variety of checks for data anomalies, such as checking for duplicates, invalid values, or outliers. DBT enables users to write custom SQL-based tests that flag data anomalies based on business logic. For example, a test might identify duplicate transaction records or check that a certain percentage of values in a column fall within an acceptable range. These tests help ensure that only high-quality data enters the data pipeline and that downstream analysis is based on reliable and accurate information.

Schema testing is another crucial part of DBT's testing capabilities. DBT can automatically verify that models conform to a predefined schema, including data types, constraints, and column names. Schema testing ensures that data models are aligned with the expected structure, which helps prevent issues during downstream processing or reporting. For example, you can define tests to confirm that a column in your data model always contains a specific data type, such as integer or string. This reduces the chances of encountering errors due to unexpected data types or missing columns when querying the data.

DBT's testing framework also includes an easy-to-use test result reporting feature, which allows users to track test failures and successes. The test results are logged and displayed in a structured format, which helps teams identify problems quickly and address them efficiently. By reviewing test results, teams can focus their attention on failing tests and determine the root cause of issues. The testing framework also integrates with DBT's documentation capabilities, providing a unified view of both the data and its quality. This comprehensive view ensures that stakeholders have visibility into both the structure and integrity of the data.

Testing can also be easily incorporated into model versioning. As DBT models evolve over time, new tests can be added to ensure that changes do not break existing functionality. Since DBT integrates well with version control systems like Git, teams can track changes to tests and models over time, ensuring that they maintain both the integrity of their transformations and the overall health of the data pipeline. Additionally, when making significant changes to a model, teams can use DBT's testing framework to verify that the new version of the model meets the same quality standards as previous versions.

The use of data snapshots in DBT also contributes to testing by capturing historical versions of data. Snapshots help teams identify when data changes unexpectedly and ensure that changes are tracked over time. This is particularly important when working with slowly changing dimensions (SCDs) in a data warehouse. DBT's snapshot feature can be used to track data changes over time, allowing teams to maintain an audit trail of data transformations and detect any anomalies that might arise due to unexpected changes in the source data.

Lastly, DBT encourages a test-first approach, which involves writing tests before developing the models. This strategy ensures that all aspects of the data transformation process are accounted for and validated from the outset. A test-first approach fosters a mindset of quality assurance and proactive data management, helping to identify potential issues early in the development cycle. By incorporating testing from the beginning, teams can deliver higher-quality data and ensure that their models continue to meet the expectations of stakeholders, providing confidence in the integrity of the data pipeline.

In conclusion, leveraging DBT's advanced testing capabilities is an essential part of building a reliable, high-quality data pipeline. From custom SQL tests and automated CI/CD integrations to schema validation and data snapshots, DBT offers a robust set of tools to ensure data integrity and accuracy. By incorporating comprehensive testing into their workflow, data teams can proactively identify issues, prevent data quality problems, and deliver trustworthy results to business users. These advanced testing features enable organizations to confidently scale their data pipelines while maintaining a high level of quality and consistency.

Managing Large-Scale DBT Projects

Managing large-scale DBT projects requires a combination of strategic planning, disciplined workflows, and robust project management practices to ensure that data transformations are scalable, efficient, and maintainable. As projects grow in complexity, data teams must tackle challenges like managing hundreds or even thousands of models, ensuring efficient collaboration across distributed teams, and maintaining consistent data quality across all stages of the pipeline. Successful management of large-scale DBT projects involves establishing clear organizational structures and best practices that support both technical and operational needs.

One of the first strategies for managing large-scale DBT projects is organizing models into logical directories and packages. A well-structured directory layout helps ensure that each model is easy to locate and maintain. For example, grouping models by business domain or data source (e.g., marketing, sales, or finance) allows teams to quickly navigate through the project and isolate relevant models. Additionally, splitting complex models into smaller, more manageable pieces improves readability and makes it easier to troubleshoot and update them as needed. Dividing work across multiple packages, each with its own directory of models and transformations, further enhances project scalability.

To keep large-scale DBT projects manageable, it's also important to prioritize modularity and reusability. Rather than creating long, complex SQL scripts for each transformation, teams should break down processes into reusable components like macros, models, and seeds. Macros allow you to write reusable SQL code that can be applied across multiple models, avoiding duplication and minimizing maintenance. Similarly, leveraging DBT's ref() function helps ensure dependencies between models are clearly defined, making it easier to track and update the transformations as the project grows. Modular components promote code reuse and simplify maintenance as the project evolves.

Collaboration becomes increasingly crucial as DBT projects scale, and effective communication between team members is essential for maintaining project consistency. Version control with Git is a cornerstone of collaboration in large-scale DBT projects, ensuring that all changes to the codebase are tracked and that multiple developers can work on different parts of the project without conflict. To further improve collaboration, teams can adopt branching strategies like feature branches, where each new feature or change is developed in isolation before being merged back into the main branch. This approach reduces the risk of conflicts and allows team members to work on different models or sections of the pipeline simultaneously.

Another challenge when managing large DBT projects is maintaining data quality and consistency across all models. In larger projects, there may be several sources of data, and maintaining the integrity of those sources is paramount. Regularly testing transformations with DBT's testing framework is an effective way to catch issues early and ensure that all models adhere to the expected standards. Additionally, having a clear governance framework in place for managing testing, quality assurance, and version control is essential to maintain consistency and prevent the introduction of errors into the pipeline.

The complexity of large-scale DBT projects also necessitates the use of automated deployment pipelines. By automating the deployment process with Continuous Integration/Continuous Deployment (CI/CD) tools like Jenkins, CircleCI, or GitHub Actions, teams can ensure that any updates to the project are deployed consistently and reliably. Automated tests should be run during each deployment to verify that changes have not broken existing functionality. The automation of model deployment allows for faster iterations, improves efficiency, and reduces the manual overhead involved in managing complex projects.

A critical consideration in large-scale DBT projects is performance optimization. As the volume of data increases, transformations can become slower, and the time it takes to run models can impact overall project performance. DBT offers several optimization techniques, such as partitioning tables, clustering data, and leveraging incremental models to only process new or changed data. These methods help reduce the computational load and improve execution times. Additionally, regularly reviewing and optimizing the performance of models ensures that they continue to scale effectively as data volumes grow.

Finally, documentation plays a vital role in large-scale DBT projects. As the number of models increases, it becomes essential to have well-maintained documentation that explains the logic behind each model and transformation. DBT's built-in documentation capabilities allow you to automatically generate documentation for all your models and their dependencies, providing a comprehensive overview of your data pipeline. Well-documented models help new team members onboard more quickly, reduce confusion, and enable effective troubleshooting when issues arise.

In conclusion, managing large-scale DBT projects requires a focus on modularity, collaboration, version control, testing, automation, performance, and documentation. By establishing transparent processes and adopting best practices, data teams can efficiently scale their DBT workflows while maintaining data quality and consistency. With the right strategies in place, teams can ensure that their DBT projects remain agile, maintainable, and capable of handling large volumes of data without compromising on performance or reliability.

Optimizing DBT Run-Time: Best Practices

Optimizing DBT run-time is crucial to ensuring that data pipelines run efficiently, especially as datasets grow larger and more complex. With the increasing volume of data, performance bottlenecks can arise in DBT models, slowing down the overall transformation process and impacting business operations. By implementing best practices for optimization, teams can reduce the time it takes to run DBT models, which ultimately leads to faster insights and better resource utilization.

One of the primary ways to optimize DBT run-time is by leveraging incremental models. Instead of re-running the entire dataset each time a model is executed, incremental models process only the new or changed data. This significantly reduces the computational load and speeds up the transformation process. To implement this, DBT allows users to specify incremental logic using the is_incremental() function, ensuring that only the necessary records are processed. By reducing the volume of data that needs to be processed, teams can dramatically improve the performance of their DBT runs.

Another key technique for optimization is the use of materializations. DBT provides different materialization strategies such as table, view, and ephemeral. Using a view materialization ensures that the model always runs the query, but it may not always be the most efficient for large datasets. Materializing models as tables or incremental tables reduces the need to recompute data repeatedly. Using the table materialization for intermediate results can improve run-time performance by persisting the results, which prevents the model from recalculating every time it's run. However, the choice of materialization depends on the specific use case and the data processing needs.

Optimizing SQL queries within DBT models is another essential practice. DBT leverages SQL as its primary query language, and inefficient SQL queries can lead to poor performance. To optimize query execution, consider simplifying SQL joins, reducing the number of subqueries, and using indexing on large tables. When working with large datasets, ensure that queries are written in a way that minimizes data movement between databases or stages. Additionally, breaking up complex SQL logic into smaller, more manageable pieces can lead to more efficient execution. Refactoring queries for simplicity and reducing unnecessary calculations can significantly improve run-time efficiency.

Partitioning and clustering tables in DBT is another effective strategy for performance optimization. Partitioning divides large tables into smaller, more manageable segments, typically based on date or another logical field. This allows for faster query execution since the database can process only the relevant partitions rather than the entire table. Similarly, clustering arranges data within partitions in a way that improves query performance by keeping frequently queried columns together. Using both partitioning and clustering can dramatically improve the performance of large-scale DBT models, particularly for analytics queries that filter on specific columns.

Optimizing the resource allocation of the underlying database is also crucial. DBT runs queries against the underlying data warehouse, and performance can be affected by the resources allocated to that warehouse. For example, some data warehouses allow for dynamic scaling, where users can increase the number of compute nodes or the size of the warehouse during peak processing times. By adjusting the resource allocation based on the required scale of data processing, teams can reduce run-time and improve the overall efficiency of the DBT models. Monitoring performance and adjusting resources accordingly are essential to ensure an optimal balance between cost and speed.

Another practice for improving DBT run-time is ensuring that dependencies are well-managed. DBT models often depend on one another, and inefficient handling of these dependencies can slow down the overall run-time. By using the ref() function in DBT, users can create explicit dependencies between models, ensuring that DBT only builds models affected by upstream data changes. This approach prevents unnecessary reprocessing of models that haven't been affected by the changes, reducing the run-time. Additionally, it's crucial to order models in a way that minimizes dependencies and avoids the need for unnecessary recalculations.

Finally, parallelization can be an essential strategy for optimizing DBT performance. DBT allows for parallel execution of models by splitting them into smaller tasks that can be run concurrently. This can be particularly useful for large datasets or complex transformations that can be broken down into smaller parts. By utilizing parallel execution, data teams can drastically reduce the time required to process multiple models and improve the efficiency of the overall pipeline. However, it's vital to ensure that the underlying data warehouse supports parallel execution and that resources are adequately allocated to handle concurrent tasks.

Incorporating these optimization strategies into DBT workflows can significantly enhance performance, reduce run-times, and improve the overall efficiency of data transformation pipelines. With the right combination of incremental models, materializations, query optimization, partitioning, and proper resource management, teams can scale their DBT projects while maintaining high performance. Regularly revisiting and fine-tuning these practices is key to adapting to growing datasets and evolving business needs.

Using DBT for Incremental Loading

Using DBT for incremental loading is a powerful technique for managing and optimizing large-scale data transformations. Incremental loading refers to processing only new or updated records in a dataset, rather than recalculating the entire dataset each time the pipeline is executed. This approach significantly reduces the amount of data that needs to be processed, improving both the performance and efficiency of data workflows. In DBT, incremental models can be configured to process only the data that has changed since the last run, which is especially beneficial when working with large data sources or frequent updates.

DBT's incremental models work by defining a set of criteria for identifying new or modified records. Typically, this is done using a timestamp or an auto-incrementing primary key. The is_incremental() function in DBT allows for conditional logic to be written into the model, ensuring that only the relevant data is included in the transformation process. This is particularly useful in scenarios where data is frequently updated, such as transaction logs or event streams. By processing only the new or modified records, DBT helps optimize resource utilization and reduce the time spent on redundant processing.

To configure incremental loading in DBT, users need to define the model as incremental by using incremental materialization. This materialization tells DBT to build the model incrementally rather than from scratch. The model will then check for changes since the last run and apply the necessary transformations to only the changed data. This allows for faster processing and reduces the computational load on the underlying database, especially when working with large datasets that would otherwise be time-consuming to process from the beginning.

One key benefit of incremental loading in DBT is that it allows for faster and more efficient model builds. Since only a small subset of data is processed, the overall run-time of the transformation is reduced. This is particularly helpful when dealing with large datasets, where the cost of rebuilding an entire table can be prohibitively high. By processing only incremental data, DBT helps streamline workflows, enabling data teams to work with up-to-date information without incurring the full cost of recalculating all data every time.

However, working with incremental loading in DBT also requires careful handling of the data pipeline to ensure data integrity. It's essential to track changes accurately and manage potential issues such as duplicate records or data anomalies. For example, when using a timestamp-based incremental approach, there might be cases where records are missed or misrepresented due to delays in data ingestion or processing. To mitigate these issues, DBT allows for fine-grained control over the incremental logic, enabling users to define thresholds, filters, and other conditions that help ensure accurate and reliable data loading.

DBT also provides several mechanisms to handle data deletions and updates that might occur during the incremental loading process. For instance, the merge strategy in DBT can ensure that records are updated or deleted correctly based on specific conditions. This is useful when a dataset needs to be updated with corrected or modified data rather than simply appending new records. By incorporating strategies like these, DBT ensures that the integrity of the dataset is maintained and that any necessary adjustments are made to reflect the most accurate information.

In addition to handling updates and deletions, incremental models in DBT can also be customized to handle different types of data. For example, in scenarios where data is partitioned by a specific key or time period, DBT allows users to configure incremental logic for each partition. This is particularly useful for time-series data or large datasets split across different segments, where each partition can be updated independently. This approach reduces the complexity of managing large-scale data pipelines and makes it easier to scale DBT models across multiple dimensions.

Lastly, monitoring and troubleshooting incremental loads is essential to ensure that the process is running smoothly. DBT provides useful logging and error-handling features to help identify issues with incremental models. For example, if an incremental load fails or produces incorrect results, DBT's logs can highlight the specific part of the model that needs attention. Regularly reviewing and maintaining incremental models, along with implementing strategies for dealing with edge cases, ensures that DBT's incremental loading capabilities can handle even the most complex data transformation scenarios.

In summary, using DBT for incremental loading helps improve data pipeline performance by reducing the amount of data that needs to be processed. By leveraging incremental materializations, users can efficiently build and maintain data models, keeping transformations fast and scalable. Through careful management of the data pipeline, including handling updates, deletions, and partitioning, DBT ensures that incremental loads remain accurate and reliable, even as data volumes grow. As a result, DBT is an invaluable tool for teams looking to optimize their data workflows and improve the overall performance of their data pipelines.

Versioning and Collaboration in Advanced DBT Projects

Versioning and collaboration are critical aspects of managing large and complex DBT (Data Build Tool) projects, especially as teams grow and work on different components of the data pipeline simultaneously. With DBT, version control and collaboration are seamlessly integrated into the development workflow, enabling teams to work together effectively while maintaining the integrity of their work. A robust versioning system, typically Git, ensures that every change is tracked, and any updates made to the models, tests, or documentation are properly versioned, so there is a clear record of what changes were made and when.

In DBT projects, versioning allows team members to keep track of changes to models and transformations, making it easier to collaborate on a shared codebase. By using Git to version control DBT projects, teams can maintain a historical record of changes, facilitate code reviews, and manage merges efficiently. Each developer can work on different branches without interfering with the main workflow, allowing for more flexible collaboration. After changes are completed and tested in a separate branch, they can be merged back into the main project, ensuring that the core project remains stable while new features are being added.

Collaboration in DBT is also supported through its integration with modern cloud platforms, such as GitHub, GitLab, and Bitbucket. These platforms offer a range of collaboration features, including pull requests, code reviews, and issue tracking. This allows teams to discuss changes, resolve conflicts, and ensure that only high-quality code is merged into the production branch. Pull requests in particular help ensure that any new transformations or updates to existing models go through an organized review process, which improves the overall quality of the DBT project.

Moreover, DBT provides an environment where data models can be developed and tested locally, which is crucial for teams working in parallel. With version control, each team member can develop changes in their local environments and push them to the central repository once they're ready. This prevents multiple developers from overwriting each other's work or making conflicting changes. It also ensures that the development process is transparent and trackable, making it easier to trace the origins of any issues or bugs in the models, should they arise.

To enhance collaboration and versioning within DBT, it is essential to follow best practices for project structure and naming conventions. Organizing DBT models, macros, and other resources in a clear and consistent manner can make it easier for collaborators to understand the structure of the project and contribute effectively. Defining a consistent naming convention for models, for example, allows developers to quickly identify their function within the project. Similarly, modularizing the project into different directories based on functionality can help improve clarity and reduce the risk of version conflicts when merging changes.

Handling dependencies is another important aspect of versioning and collaboration. DBT projects often involve multiple models and sources, and understanding their interdependencies is crucial when making changes. DBT helps manage these dependencies by allowing users to explicitly define relationships between models using the ref() function, which ensures that models are built in the correct order. In complex projects, where different models rely on each other, version control also helps track which versions of models were used in previous runs, making it easier to roll back to an earlier version if needed.

Effective collaboration with DBT also extends to managing data documentation and testing. As teams develop new models, it's essential to document them properly for future reference and to ensure that others can understand and maintain them. DBT makes it easy to add descriptions and documentation for models, columns, and tests directly within the project files. Versioning these documentation updates alongside the code itself ensures that the entire team is constantly working with the most up-to-date information. Similarly, testing can be versioned and integrated into the collaboration process, helping to catch issues early and ensuring that each new change meets the project's quality standards.

Finally, advanced DBT projects often involve larger teams with multiple contributors working on various aspects of the pipeline. To ensure smooth collaboration at scale, teams can implement policies around branch management, code reviews, and testing. For example, teams may enforce a policy where changes to critical models or transformations are only merged after passing specific tests or after being reviewed by a senior team member. This ensures that the project remains stable and maintainable as it grows.

In summary, versioning and collaboration in DBT are vital components of managing large and complex data transformation projects. By using version control systems like Git and integrating them with platforms such as GitHub, teams can maintain a clear and organized development workflow. Clear naming conventions, modular project structures, and proper dependency management further enhance collaboration, ensuring that changes are made efficiently and conflicts are minimized. Ultimately, the combination of DBT's built-in versioning capabilities and external tools for collaboration empowers teams to work together effectively, maintain high-quality code, and deliver reliable data transformations at scale.

DBT Artifacts: Understanding Logs and Artifacts for Debugging

DBT (Data Build Tool) generates various artifacts during the execution of models, tests, and other tasks. These artifacts are essential for debugging, performance monitoring, and understanding how DBT executes a project. Understanding the logs and artifacts

DBT produces allows data engineers and developers to identify issues, track down failures, and optimize the data transformation pipeline. DBT provides a well-structured and detailed set of logs that record the entire run, capturing errors, warnings, execution times, and even the results of tests. These logs are key to troubleshooting problems and improving the overall performance of the models.

The primary artifact generated by DBT is the run artifact, which contains a detailed log of each executed model. This log captures the execution status of each model, including whether it succeeded, failed, or was skipped. It also includes the time taken for each model to execute, providing helpful performance metrics. If a model fails, DBT logs the specific error message associated with that failure, which can be invaluable in pinpointing the issue. These logs are written to the command line output by default but can also be saved as a file for later analysis, making it easy for teams to revisit issues and debug them in future iterations.

Another key artifact is the DBT manifest.json file, which is automatically generated at the end of every run. This file contains metadata about the DBT project, including information on models, sources, and tests, as well as their dependencies. It provides a comprehensive view of the project's structure and can be used to understand the relationships between different models and sources. The manifest is especially useful for debugging dependency-related issues, as it tracks the order in which models should be run and which models rely on others. By analyzing the manifest, developers can determine whether DBT is executing models in the correct order or if there are any discrepancies in the model dependencies.

In addition to the manifest.json, DBT generates a run_results.json file, which contains a record of each model's execution results. This artifact includes the status of each model (success, failure, or skipped), the number of rows processed, and other key details about the execution. The run_results.json is valuable for automated monitoring and alerting, as it enables teams to track the health and success of their data pipelines in real time. When something goes wrong, the run_results.json file can quickly provide insight into where the failure occurred, helping developers to focus on the most relevant models or transformations that need attention.

DBT also produces detailed logs for the execution of tests within the project. These logs are captured in the test_results.json file, which provides information on the success or failure of each test, including the specific assertions made during the testing process. If a test fails, the logs capture the exact reason for the failure, which can include mismatches in expected versus actual results, missing data, or other issues. These test logs are crucial for maintaining the integrity of data models and ensuring that they meet the desired quality standards. By reviewing the test logs, developers can identify which parts of the transformation pipeline need to be addressed to resolve issues and improve the overall reliability of the models.

For debugging complex issues, DBT provides additional logging levels that can be configured to provide more detailed insights into the execution process. By setting the logging level to debug, DBT outputs additional details that can help pinpoint the exact cause of issues. This includes detailed SQL queries being executed, as well as any errors that occur within the database during the execution of the models. The debug logs are especially useful when the issue is not easily identifiable through the default logs. These more granular logs can be helpful when working with complex data transformations or when the problem is specific to the underlying database.

Another valuable tool for debugging DBT models is the dbt debug command, which performs a set of diagnostic checks on the DBT environment. This command checks the configuration of the DBT project, the database connection settings, and other key components to ensure that everything is set up correctly. If there are issues with the DBT project or the connection to the data warehouse, the dbt debug command provides actionable insights and recommendations for resolving these issues. This makes it an invaluable tool when encountering problems that are not related to the logic of the models but instead to the environment or configuration.

DBT also supports the integration of external tools for advanced debugging and monitoring. For example, by integrating DBT with tools like Datadog or Sentry, teams can receive more granular alerts and performance metrics related to DBT execution. These integrations allow for monitoring DBT runs in real time, catching errors as they occur, and providing detailed information on what went wrong. This can be particularly useful in large-scale projects where manual log checking can be time-consuming. Automated error detection and monitoring ensure that any issues are caught early, minimizing the impact of failures on the overall data pipeline.

In summary, DBT artifacts, such as logs, manifest files, run results, and test results, are critical for debugging and performance optimization in data transformation pipelines. These artifacts provide detailed insights into the execution of models, tests, and other tasks, helping teams track down issues, optimize performance, and maintain the integrity of their data pipelines. By leveraging DBT's built-in logging capabilities, using tools like dbt debug, and integrating with external monitoring systems, developers can ensure that their DBT projects run smoothly and efficiently, minimizing downtime and improving data quality.

Scaling DBT for Enterprise-Level Deployments

Scaling DBT for enterprise-level deployments requires careful planning and a robust architecture to handle the increased complexity and volume of data transformations. As organizations grow and the amount of data and models increases, DBT workflows must evolve to maintain performance, reliability, and collaboration across large teams. One of the first considerations in scaling DBT is optimizing the infrastructure. Enterprises often leverage cloud-based data warehouses like Snowflake, BigQuery, or Redshift, which provide the scalability and performance necessary to handle large-scale data transformations. Organizations can efficiently manage the growing demands of an expanding data pipeline by tuning DBT's configuration to work seamlessly with these cloud data platforms.

When scaling DBT, managing dependencies becomes crucial. In large DBT projects, there are often complex relationships between models, sources, and tests. Without proper dependency management, there can be issues with execution order, model failures, and data inconsistencies. DBT's ability to manage these dependencies automatically helps in large-scale deployments, but organizations must also adopt strategies such as modularization and incremental loading. Modularizing the models allows teams to build, test, and deploy smaller, more manageable components, which can then be recombined to form the larger pipeline. Additionally, enabling incremental loading for models reduces the amount of data that needs to be processed, thus improving performance and reducing processing time.

Collaboration among multiple teams is another challenge when scaling DBT for enterprise deployments. Large organizations often have multiple teams working on different parts of the data pipeline, from data engineering to analytics and data science. To facilitate collaboration, it is important to adopt a version control system like Git to manage changes to DBT models, configurations, and tests. Organizations can ensure that team members can work on different parts of the pipeline concurrently without conflicts by setting up proper branching strategies and merge workflows. DBT's ability to integrate with Git allows teams to manage their work in a collaborative manner and track changes over time, helping to avoid issues like model conflicts and untested changes.

For enterprise-level deployments, monitoring and alerting systems become critical for ensuring the reliability of DBT workflows. As the data pipeline scales, it becomes more difficult to manually check every transformation. Therefore, setting up automated monitoring through external tools like Datadog, Sentry, or Airflow is essential. These tools can track the success or failure of DBT runs, monitor resource usage, and send alerts when performance thresholds are exceeded. Monitoring ensures that potential issues are caught early, minimizing the impact of failures on downstream analytics and business processes. It also provides valuable insights into the performance of the entire data pipeline, helping teams optimize DBT models and reduce bottlenecks.

Performance optimization also plays a key role in scaling DBT for enterprise deployments. As data volumes grow, it's important to fine-tune DBT's performance to minimize the time taken for models to run and reduce resource consumption. Leveraging DBT's incremental models, which only update new or changed data, can significantly reduce run-times. Additionally, optimizing SQL queries for performance, parallelizing tasks, and utilizing data warehouse features such as clustering or partitioning can further improve processing times. Enterprises must regularly review and adjust DBT configurations to ensure that their data pipeline continues to perform efficiently as the scale of data grows.

Finally, enterprises must consider the long-term management of their DBT projects. As the project grows in size, keeping it organized and maintainable becomes essential for long-term success. This involves setting up clear guidelines for model naming conventions, testing strategies, and version control workflows. It also requires maintaining comprehensive documentation for models, sources, and transformations so that future team members can quickly understand and contribute to the project. By establishing best practices and governance policies around DBT, enterprises can ensure that their data transformations remain manageable, efficient, and aligned with business goals even as the project scales over time.

Conclusion

As organizations move toward enterprise-level data projects, scaling DBT effectively becomes a key factor in maintaining performance, collaboration, and consistency across large datasets and teams. By optimizing infrastructure, managing dependencies, and using version control systems, enterprises can create a robust and scalable data pipeline that supports the needs of their growing business. Emphasizing best practices in collaboration, testing, and monitoring ensures that DBT projects are not only scalable but also reliable and maintainable in the long run.

Incorporating DBT's advanced features, such as incremental loading, performance optimization techniques, and clear documentation, provides enterprises with the tools necessary to handle the complexity of large-scale deployments. By focusing on these strategies, organizations can confidently scale their DBT pipelines, ultimately improving data accessibility, quality, and decision-making processes across the enterprise. The key to success lies in balancing performance, collaboration, and governance while ensuring that the data pipeline evolves alongside business needs and technological advancements.

CHAPTER 11

Introduction to the DataOps.live Platform

The DataOps.live platform is designed to streamline and enhance the implementation of DataOps practices within organizations. As data environments become more complex and dynamic, the need for a solution that can effectively manage data workflows, automate processes, and promote cross-functional collaboration has never been more critical. This chapter will introduce the core features and capabilities of the DataOps.live platform, providing a foundational understanding of how it can be utilized to optimize data operations, ensure data quality, and accelerate data delivery across the entire data lifecycle.

By focusing on automation, continuous integration, and real-time monitoring, DataOps.live simplifies the management of data pipelines, reducing manual intervention and increasing efficiency. The platform's ability to integrate seamlessly with various data tools and cloud environments allows teams to maintain agility while ensuring that data processes are both scalable and compliant with governance requirements. This chapter will explore how DataOps.live supports key DataOps principles, enabling organizations to overcome common data challenges and drive better business outcomes through more reliable and efficient data management.

About the DataOps.live Platform

The DataOps.live platform is a comprehensive solution designed to accelerate and optimize the entire data lifecycle. It provides an integrated environment where organizations can manage data workflows, automate tasks, and enhance collaboration among cross-functional teams. By adopting a DataOps approach, which is rooted in agile methodologies, the platform empowers data engineers, analysts, and operations teams to work together seamlessly, ensuring a more efficient and reliable data pipeline

R. L. Steelman, Jr., *Mastering Snowflake DataOps with DataOps.live*, https://doi.org/10.1007/979-8-8688-1754-0_11

management process. DataOps.live offers end-to-end data pipeline orchestration, real-time monitoring, and automated testing, helping teams to maintain consistent data quality while reducing the time-to-insight.

At its core, DataOps.live supports continuous integration and continuous delivery (CI/CD) practices, ensuring that data workflows are continuously tested, validated, and deployed with minimal manual intervention. This leads to faster development cycles, fewer errors, and improved operational efficiency. The platform's ability to integrate with a wide range of data sources, tools, and cloud environments allows for a flexible and scalable solution that adapts to the needs of any organization, regardless of size or complexity. It enables automated data ingestion, transformation, and analytics while providing a unified view of all activities in the data pipeline.

In addition to its core functionalities, DataOps.live also provides robust data governance features that ensure compliance with industry standards and regulations. The platform's security model allows organizations to control access to sensitive data and ensure that data workflows adhere to best practices in governance. It provides detailed logging and auditing capabilities, which are critical for tracking changes and ensuring accountability across the entire data pipeline. This level of visibility and control is crucial for organizations that need to manage complex data environments while mitigating risks associated with data security and privacy.

DataOps.live's user-friendly interface and powerful automation capabilities simplify the complexities of modern data management, enabling organizations to focus on leveraging data for strategic decision-making. By fostering collaboration, automating repetitive tasks, and maintaining high standards of data quality and security, the platform helps organizations unlock the full potential of their data assets, driving better business outcomes and supporting data-driven innovation.

Key Features and Capabilities

The DataOps.live platform is designed to accelerate data engineering workflows by providing a robust environment for building, testing, and deploying data products. One of the key features of DataOps.live is its ability to help teams quickly create fully governed and documented MVPs in under 10 minutes, streamlining the process of developing Snowflake Native Apps, integrating dbt Core projects, and adhering to DevOps and CI/CD best practices. This functionality is particularly beneficial for teams that want to focus on rapid development and ensure their data pipelines are trustworthy and production-ready.

The platform is not just about creating data products quickly but also about amplifying collaboration and productivity across data teams. DataOps.live integrates a powerful AI-powered assistant, "Assist," which acts as a copilot for every team member, making it easier to handle tasks like data generation, model construction, and documentation. This assistant helps to simplify complex tasks, such as summarizing intricate SQL queries, and accelerates the process of resolving pipeline failures, ensuring teams can maintain their focus on business outcomes rather than technical roadblocks.

In addition to assisting with development and collaboration, DataOps.live excels in orchestration and governance, enabling teams to manage data products throughout their lifecycle. Whether building, testing, or deploying data solutions, the platform ensures a seamless workflow, automating critical tasks such as pipeline orchestration and change management while preserving strict governance. By leveraging its end-to-end capabilities, data teams can confidently release data applications into production with minimal risk, allowing them to maintain agility while ensuring compliance with industry standards.

With DataOps.live, data engineers and analysts can also benefit from integrated observability and comprehensive logging features. These tools provide visibility into data products and pipelines, making it easier to monitor performance, troubleshoot issues, and maintain overall system health. The platform also offers tools for federated governance, ensuring that teams can manage their data across multiple environments while maintaining control over access, security, and compliance.

Overall, DataOps.live is a comprehensive solution designed to transform how organizations approach data engineering. It simplifies the development process, accelerates the time-to-market for data products, and fosters a collaborative environment where teams can work together more effectively. Through a combination of AI-powered assistance, seamless integration with Snowflake, and robust governance features, DataOps.live is poised to be a game-changer for organizations looking to optimize their data operations.

Development Principles

DataOps.live supports a streamlined, structured approach for deploying DataOps projects, and this structure is highly recommended for setting up new projects within the platform. At the highest level, the standard project structure includes three primary folders—dataops, pipelines, and vault-content—along with a CI YAML file that defines pipeline configurations. This organized structure ensures that all project components are systematically categorized, making it easier to manage, scale, and maintain the project.

The dataops directory plays a central role in the project setup. It contains key subdirectories, such as modelling, snowflake, and profiles. The modelling subdirectory is powered by DataOps.live's Modelling and Transformation Engine (MATE), where all DBT project files are located, including definitions for sources, models, seeds, tests, and various DBT configurations. MATE enhances the modelling process by offering powerful transformation capabilities to handle the intricacies of data engineering tasks.

The snowflake subdirectory is where users interact with the Snowflake Object Lifecycle Engine (SOLE), a critical feature of the DataOps.live platform. SOLE provides seamless integration with Snowflake, enabling automated management of your Snowflake objects, which simplifies the typically complex process of managing Snowflake configurations. This functionality sets DataOps.live apart from standard DBT deployments, providing a more comprehensive and integrated data operations experience.

Within the dataops directory, the optional profiles subdirectory houses custom DBT profiles that can be leveraged by MATE. If no custom profiles are defined, the platform defaults to a standard DBT profile, ensuring that projects continue to function even without custom configurations. This level of flexibility allows teams to tailor the DBT environment to meet specific needs while maintaining compatibility with the broader DataOps.live platform.

The pipelines directory contains all configurations related to pipeline execution. It contains essential files like agent_tag.yml, which specifies the DataOps Runner for project pipeline jobs; stages.yml, which customizes stage names within the project; and variables.yml, where key configuration variables are set. These variables often include settings related to debugging, Snowflake warehouses, GIT collaboration branch names, and run-time script execution. By using these configuration files, teams can effectively manage the behavior of their data pipelines.

In addition to the primary configuration files, the pipelines directory also features a local_includes subdirectory for organizing job definitions. This allows for better management of job configurations in complex projects and facilitates more straightforward navigation. For example, a modelling_and_transformation subdirectory could house all MATE-related tests, transformations, and documentation generation jobs, further organizing the workflow and simplifying maintenance.

Finally, the vault-content directory usually contains a single file—vault.template.yml—that defines additional vault configurations. This file is crucial for managing sensitive values, as it integrates with the DataOps Runner's vault files or configured secrets managers, ensuring that confidential data is securely stored and accessed as needed. This integration of vault functionality adds an extra layer of security, particularly in projects dealing with sensitive or regulated data.

Use Cases: DataOps.live in Snowflake Data Cloud

DataOps.live enables streamlined management and orchestration of data pipelines within the Snowflake Data Cloud, allowing teams to enhance their data workflows. It facilitates the entire data product lifecycle, from creation to deployment, ensuring governance and agility in production environments. Through automated testing, CI/CD integration, and robust observability, teams can maintain secure and high-quality data operations.

The platform's focus on Snowflake-native applications, including UDFs, AI/ML models, and Snowpark services, ensures seamless integration with Snowflake, enhancing the platform's capabilities. Teams can rapidly develop Snowflake-native apps for the Snowflake Marketplace, promoting efficiency and governance from the start.

DataOps.live integrates tightly with dbt Core™, enabling smooth data transformations while maintaining complete DevOps and CI/CD support. This integration empowers teams to automate their workflows, simplifying the management of complex data environments, improving the speed of development, and ensuring a more reliable release management process.

Governance is a primary pillar in DataOps.live's approach ensures that each data product is not only compliant but also aligned with business objectives. This is especially valuable in industries where data accuracy and compliance are paramount. The platform helps bridge the gap between data engineers and business stakeholders by offering full transparency throughout the data development process.

With its powerful AI-powered assist features, DataOps.live empowers users to accelerate their productivity. Automating tasks such as model creation, test generation, and data summarization reduces manual effort and errors, allowing teams to focus on solving business problems. The assist function also streamlines incident resolution, shortening the time to address issues in data pipelines.

Another important use case for DataOps.live is its ability to foster collaboration among team members. The platform supports multiple roles, including developers, data engineers, and business stakeholders, helping teams work together more efficiently. Features such as change tracking, peer reviews, and automated documentation enhance communication and ensure that everyone is on the same page.

The real-time observability features of DataOps.live provides teams with comprehensive monitoring capabilities, enabling them to track data flow, identify issues early, and respond promptly. This results in reduced downtime and improved reliability, which is critical for organizations that rely on data-driven decisions in real time.

Conclusion

The DataOps.live platform represents a significant advancement in managing data pipelines within the Snowflake Data Cloud. By offering seamless integrations, robust governance, and a comprehensive set of tools, it empowers teams to automate processes, optimize performance, and collaborate more effectively. These capabilities are essential for businesses looking to unlock the full potential of their data infrastructure while maintaining high standards of compliance and reliability.

As organizations continue to rely on data-driven decisions, the need for platforms like DataOps.live becomes increasingly apparent. Its emphasis on automation, observability, and collaboration makes it a powerful tool for data teams looking to enhance productivity and ensure that data flows seamlessly from development to deployment.

CHAPTER 12

DataOps.live and DataOps: Better Together

The combination of DataOps principles with a robust platform like DataOps.live offers decisive advantages for organizations striving to manage complex data environments. DataOps as a concept emphasizes the need for agility, collaboration, and automation in the data lifecycle, and when implemented effectively within a platform, it accelerates time-to-insight while ensuring data governance and quality. In this chapter, we explore how integrating DataOps principles into a dedicated platform creates synergies that enhance data management workflows and deliver scalable, high-performance solutions.

By leveraging the capabilities of the DataOps platform, teams can streamline their workflows, from development to deployment, with enhanced automation, monitoring, and collaboration. This chapter will discuss the core aspects of the platform that elevate traditional DataOps practices, providing an in-depth look at how the integration of advanced tools and features supports these principles in a way that maximizes the efficiency, security, and scalability of data operations across various industries.

Integration of DataOps Principles into DataOps.live

Integrating DataOps principles into the DataOps platform is fundamental to the overall success of data management workflows. The platform has been designed with the core principles of DataOps in mind: automation, collaboration, and agility. One of the key ways this integration manifests is through automation of manual processes that traditionally slow down data development cycles. By embedding continuous integration and continuous deployment (CI/CD) practices, the platform ensures that changes to data pipelines and models are tested, reviewed, and deployed seamlessly. This process reduces human errors and accelerates delivery cycles, aligning with the agile methodology central to DataOps.

R. L. Steelman, Jr., *Mastering Snowflake DataOps with DataOps.live*, https://doi.org/10.1007/979-8-8688-1754-0_12

Another crucial element in integrating DataOps principles is the focus on collaboration. In the DataOps platform, users can work together more effectively by accessing shared repositories, maintaining version control, and fostering transparent communication through easily trackable change logs. The platform's collaborative environment supports cross-functional teams, from data engineers to analysts, by providing tools that streamline communication and ensure that all team members are working from the most up-to-date datasets and models. This environment of collaboration also promotes faster decision-making and aligns teams toward a unified data strategy.

Data governance is another core principle of DataOps that is seamlessly incorporated into the platform. The platform provides mechanisms for enforcing consistent data quality, security, and compliance across the entire data lifecycle. With built-in validation rules, automated data quality checks, and integrated audit trails, organizations can ensure that their data not only meets regulatory standards but also maintains integrity throughout its life cycle. This level of governance is critical for organizations that handle sensitive data and need to mitigate risks associated with data breaches or inconsistencies.

Furthermore, the platform's ability to integrate with various data storage and processing systems, such as cloud environments like Snowflake, enables teams to apply DataOps principles at scale. This scalability allows organizations to manage larger datasets, higher complexities, and more diverse data sources without sacrificing performance or governance. By supporting a wide range of data sources, the platform ensures that organizations can apply DataOps practices to their entire data ecosystem, streamlining workflows and reducing silos that often hinder collaboration.

Additionally, the DataOps platform provides a unified view of all data pipelines, metrics, and operational logs. This visibility is key to maintaining a high level of control over data operations. With the help of comprehensive logs and reporting tools, teams can easily monitor pipeline performance, detect issues early, and troubleshoot. This enhanced visibility fosters a proactive approach to maintaining data health, enabling teams to detect and resolve issues before they impact the business.

The integration of testing and monitoring frameworks further enhances the DataOps platform. By automating testing procedures, the platform ensures that data models and pipelines are continuously validated, meeting performance and quality expectations. Moreover, the platform's ability to monitor data pipelines in real time allows for immediate detection of failures or deviations, providing teams with actionable insights. This automated feedback loop contributes to faster iterations and ensures that data processes remain aligned with business goals.

Finally, integrating DataOps principles into the platform isn't just about the tools—it's about fostering a culture of data-driven decision-making and continuous improvement. The platform's capabilities empower teams to iterate on their processes and refine their approach, aligning data initiatives with the broader business objectives. This cultural shift, alongside the technological capabilities of the DataOps platform, helps organizations embrace the full potential of their data and adapt more quickly to evolving needs.

In summary, the integration of DataOps principles into the platform isn't just about technology—it's about creating a robust framework that enhances collaboration, automation, governance, and scalability. By embedding these principles, the DataOps platform enables teams to work more efficiently, deliver higher-quality results, and continuously improve their data management practices. This holistic approach is critical for organizations that want to unlock the full value of their data in a rapidly changing digital landscape.

How DataOps.live Enhances DataOps Practices

DataOps.live enhances DataOps practices by providing a unified platform that streamlines data pipeline management, ensuring efficient data flow from source to consumption. The platform's core strength lies in automating repetitive tasks, such as testing, deployment, and monitoring. This reduces human error, speeds up development cycles, and promotes consistency across data pipelines. By automating these processes, teams can focus on higher-value tasks like analyzing data and improving data models.

Collaboration is another critical area where DataOps.live shines. By fostering a collaborative environment where cross-functional teams can access and contribute to the same data workflows, the platform breaks down silos that typically hinder data operations. Whether it's data engineers, analysts, or business stakeholders, everyone is on the same page, which improves transparency, accelerates decision-making, and promotes a culture of shared responsibility for data quality.

The platform also enhances DataOps practices by integrating comprehensive monitoring and feedback loops. With real-time visibility into pipeline performance and automatic error detection, DataOps.live ensures teams are always aware of the health of their data pipelines. This proactive monitoring allows for quicker resolutions of issues, prevents data-related bottlenecks, and enables continuous delivery of high-quality data.

DataOps.live's integration of security and governance within its workflows adds another layer of benefit. By embedding security and compliance checks directly into the data pipeline, organizations can safeguard data and ensure regulatory compliance without requiring additional manual oversight. This end-to-end governance framework ensures that sensitive data is protected and that teams adhere to industry standards, making it easier to manage risk in complex data environments.

Another key enhancement to DataOps practices is the platform's ability to scale seamlessly. Whether an organization is managing a small data environment or scaling to handle big data, DataOps.live supports scalability without compromising performance or security. This is critical for organizations that anticipate significant data growth and need a platform that can adapt to their evolving needs.

Automation also plays a central role in improving the quality and efficiency of data operations. DataOps.live provides robust automation features, including continuous integration/continuous delivery (CI/CD), automated testing, and seamless deployment of data models. This level of automation ensures that teams can roll out new updates and features rapidly while maintaining a consistent quality and reducing downtime, ultimately accelerating time-to-value for data initiatives.

Finally, DataOps.live enhances DataOps practices by offering flexibility and integration with other tools and systems. Its compatibility with various cloud environments, databases, and data warehouses like Snowflake ensures that teams can seamlessly integrate their existing workflows into the platform by supporting a wide range of integrations, DataOps.live enables organizations to bring their data from diverse sources into a unified workflow, improving efficiency and accelerating their ability to deliver data-driven insights.

Through these features, DataOps.live streamlines data operations and enables organizations to continuously evolve and optimize their data practices. Providing a platform that integrates automation, collaboration, governance, and scalability empowers teams to operate with agility, responsiveness, and reliability, key tenets of successful DataOps.

Case Studies: Illustrating the Synergy Between DataOps and DataOps.live

A leading solar energy company faced the complex challenge of efficiently managing vast amounts of data, ranging from energy production metrics to environmental data gathered from multiple sources such as solar panels, weather stations, and energy grids.

The company was looking to improve its data pipeline efficiency, enhance data quality, and ensure fast access to actionable insights. To address these needs, they turned to DataOps principles and the DataOps.live platform to streamline and modernize their data operations. Snowflake and the Medallion Architecture formed the backbone of their solution, providing a solid foundation for efficient data processing and storage.

The company's data strategy was built on a three-layer architecture inspired by the Medallion approach: Bronze, Silver, and Gold layers. The first layer, the Bronze layer, acted as the foundational data lake. Data was ingested from various sources using MuleSoft, which facilitated the integration of disparate systems. The data lake served as the initial repository, holding raw, untransformed data from solar production systems, environmental sensors, and operational logs. This data needed to be organized and structured for further use, but it was essential for the company to have a centralized location to store all incoming raw data.

The Silver layer of the architecture used a Data Vault 2.0 model, which was implemented using DataOps.live in conjunction with dbtVault (now Automate DV). This layer transformed the raw data into a more structured, usable form. The Data Vault 2.0 model provided a robust approach to data warehousing, enabling the company to capture historical data, ensure auditability, and maintain flexibility as new data sources were added. DataOps.live facilitated the automation of data pipeline creation, management, and testing, ensuring that the transformation and loading processes were seamless, repeatable, and auditable. This allowed the data engineering team to focus on more value-added tasks, such as optimizing data models, rather than getting bogged down in manual pipeline management.

Once the data had been processed and transformed in the Silver layer, it moved into the Gold layer, which served as the data mart. The company implemented the Kimball methodology to design data models supporting business intelligence (BI) and analytics. The Gold layer stored the most refined data, organized by subject areas like energy consumption, solar panel efficiency, and predictive maintenance. By leveraging DataOps.live, the company ensured that these data marts were updated in real-time, allowing business users to quickly query the data and generate insights without waiting for traditional batch processing cycles.

The key challenge the company faced in the early stages of the project was ensuring that each layer was integrated correctly and that data flowed seamlessly between them. DataOps.live provided them with the tools to automate the entire data pipeline, from the ingestion in the Bronze layer through the transformations in the Silver layer to the final reporting layer in the Gold layer. This sped up the process and ensured that data governance and compliance were maintained across the entire pipeline, which was critical for an energy company operating in a highly regulated industry.

Automation was a central tenet of the company's transformation. With DataOps.live, the company was able to set up automated workflows for testing, deployment, and monitoring, improving the reliability and performance of their data pipelines. The integration with dbtVault enabled automated transformations using the Data Vault 2.0 methodology, while Snowflake's scalability and performance ensured that the large volumes of data could be processed efficiently. The combination of these technologies helped the company build a flexible, scalable, and high-performance data platform that could grow alongside its expanding needs.

Furthermore, DataOps.live provided powerful features for collaboration across teams. Data engineers, analysts, and business users could work within the same platform, with clear visibility into the data flows, pipeline configurations, and transformation processes. This was particularly important in an organization where multiple departments needed to interact with the data at various stages, from raw ingestion to final reports. The platform enabled real-time collaboration and communication, which helped ensure that everyone involved in the project was aligned and that changes to data models were managed correctly.

One of the most significant improvements came from better decision-making capabilities. The solar energy company had previously struggled to generate real-time insights from its data. With the streamlined, automated pipeline and real-time data access in the Gold layer, business teams could leverage up-to-date insights on solar production, operational efficiency, and energy consumption trends. This allowed them to make data-driven decisions more quickly and effectively, such as optimizing energy distribution or identifying underperforming assets.

Another notable outcome was the company's ability to scale its data operations. As the solar energy business grew, the amount of data generated by the solar panels, weather systems, and grid monitoring grew exponentially. Snowflake's scalability, combined with the modularity provided by DataOps.live meant that the company could quickly expand its data platform to handle increased data loads without worrying about performance bottlenecks or downtime. This scalability was critical as the company expanded into new geographic regions and added more solar assets to its portfolio.

DataOps.live also played a critical role in maintaining data security and compliance. Given the nature of the energy sector and the sensitive data involved, the company needed to follow strict governance policies. DataOps.live provided automated monitoring, data lineage tracking, and auditability across the data pipelines, ensuring that data was handled securely and according to industry standards. This level of transparency helped the company maintain trust with regulators, stakeholders, and customers.

The success of this project demonstrates how DataOps principles, when coupled with a robust platform like DataOps.live, can transform data operations for companies in the solar energy sector. By adopting a Medallion Architecture, integrating Snowflake, and automating their data pipelines, the company could scale its operations, improve data quality, and enable faster, more informed decision-making. The synergy between DataOps and DataOps.live allowed the company to maximize the value of its data while maintaining compliance, governance, and security.

Conclusion

This case study illustrates the powerful impact of combining DataOps principles with a platform like DataOps.live in a solar energy context. The platform's ability to automate data workflows, ensure governance, and support scalability made it an invaluable tool for the company as it navigated the complexities of managing vast amounts of solar energy data. By leveraging the Medallion Architecture, Data Vault 2.0, and Kimball methodology, the company was able to build a robust, scalable, and efficient data ecosystem that improved decision-making and operational efficiency.

CHAPTER 13

Essential Elements of DataOps.live

This chapter will explore the essential elements that make up the DataOps.live platform, providing the building blocks for efficient and scalable data operations. This includes a deep dive into the core components such as accounts, groups, subgroups, and projects, which form the foundation of your DataOps.live environment. Additionally, we will cover pipelines and jobs, discussing how to configure and structure your projects for optimal performance.

Understanding how to adjust your project settings effectively ensures that your data pipelines are running smoothly. The following sections will provide practical insights into organizing and managing your DataOps.live projects to maximize efficiency, streamline workflows, and align with best practices.

Accounts

Accounts in DataOps.live define the organizational framework within which a client operates. A DataOps account acts as a customer tenant, the highest level in the structure. Accounts contain groups and subgroups, where projects are organized. Each project can be considered a repository containing configurations and code, similar to version-controlled platforms. A project can be created at the account level, but it is recommended to develop projects within groups or subgroups for better management and scalability.

Each DataOps.live account can have multiple users, who are granted access to the account and its associated groups and projects. Access is governed through various permission levels, ensuring that users can work within specific scopes of the account. This user management structure helps define roles and ensures that team members are aligned with organizational objectives, whether managing infrastructure or implementing DataOps pipelines.

R. L. Steelman, Jr., *Mastering Snowflake DataOps with DataOps.live*, https://doi.org/10.1007/979-8-8688-1754-0_13

Furthermore, DataOps.live allows for user role assignments, where members can be granted different access levels, ensuring proper governance and security protocols are followed. As projects grow, this tiered access structure becomes critical in maintaining smooth operations and collaboration. With this structured approach, DataOps.live effectively supports a wide range of users, from developers to administrators, ensuring everyone can contribute efficiently while adhering to governance policies.

Groups and Subgroups

In DataOps.live, groups and subgroups are essential organizational components within an account, providing a structured way to manage data projects and related resources. These concepts allow for better separation and allocation of tasks, responsibilities, and access control. Groups can house several projects and subgroups, organizing resources hierarchically that aligns with the company's data operations needs.

The ability to assign different privilege levels to members within these groups enhances collaboration while maintaining security. Group owners can delegate specific permissions to ensure that the right team members can access the necessary resources. This structure is particularly valuable for managing large-scale data operations across different teams or departments.

Furthermore, grouping data projects within subgroups fosters a more organized and scalable environment for growing data teams. Subgroups can be used to separate different aspects of a project or data domain, ensuring a clear delineation of responsibilities. This can also support varying access levels for sensitive data and projects, with higher security settings applied to specific subgroups.

Groups and subgroups help streamline the development process in DataOps.live by maintaining a structured environment. For instance, when setting up multiple DataOps pipelines, the grouping mechanism ensures that the right people manage different pipeline stages. This also reduces the risk of errors by ensuring that each team can focus on the components relevant to their responsibilities without interference from other teams.

Groups and subgroups promote efficiency in addition to organization. Linking a group with a runner simplifies computing resource management. Runners, lightweight containers executing pipeline tasks, can be registered to a specific group rather than individual projects, further optimizing resource allocation and tracking.

Groups also play a critical role in governance and compliance. Organizing data projects under specific groups makes it easier to implement rules and track data lineage, ensuring that data governance standards are adhered to. This helps teams maintain accountability and transparency across the organization's data ecosystem.

In large organizations, where multiple teams may be involved in the data lifecycle, groups and subgroups are the backbone of a collaborative environment. Teams can work independently but still adhere to an overarching strategy, facilitating cross-functional cooperation while maintaining clear lines of responsibility and access. This flexibility enables both agility and control, which are key principles in the DataOps methodology.

Overall, groups and subgroups within DataOps.live enhance both operational and strategic alignment by improving collaboration, access control, and governance across data projects. Their importance lies not only in how they organize but also in how they enable effective, secure, and scalable data operations within the platform.

Projects

In DataOps.live, the project is a key structural unit, the backbone of any development initiative. Projects in DataOps.live allow teams to organize, manage, and execute workflows in a highly structured manner. A project is a repository containing configurations, code, and data necessary to perform specific tasks, such as ETL (Extract, Transform, Load) or data analysis processes. By grouping related tasks, projects offer a centralized point for managing the lifecycle of data operations, making it easier to scale and maintain complex data workflows.

One significant benefit of using Projects in DataOps.live is their inherent flexibility. The platform allows users to structure their projects according to the organization's needs or specific use cases. For instance, data engineering teams can have distinct projects for staging, transformation, and production tasks, each isolated but integrated within a broader ecosystem. This separation helps manage dependencies more effectively, reduce operational risks, and streamline collaboration across teams, as different groups can work on various projects without interfering with one another's work.

Another key advantage is the Git-compliant nature of DataOps.live Projects. Each project behaves similarly to a Git repository, where version control becomes a natural aspect of the development process. Changes to code, configurations, or pipelines are tracked and can be merged or reverted as needed. This integration with version control ensures that project teams can collaborate without fear of overwriting each other's work or losing valuable historical context. Moreover, maintaining multiple versions of a project simultaneously enables teams to test different features or updates without impacting ongoing production work.

In addition to version control, DataOps.live Projects offer robust integrations with other tools and platforms, enabling seamless execution of data pipelines. This level of integration is crucial for organizations that operate in dynamic environments where external tools like Snowflake, AWS, or Azure play a central role. DataOps.live allows configuring projects to interact with these platforms directly, streamlining workflows that might otherwise require extensive manual intervention.

The organization and structure of DataOps.live Projects also provide significant visibility and governance. By setting clear parameters within a project, teams can better control the data flow through the pipeline. This not only enhances the process's efficiency but also improves transparency. Administrators can easily track which users have made changes to the project, review which pipeline stages have been executed, and monitor the overall health of the data workflows. This built-in governance helps ensure compliance with organizational policies and industry standards.

Furthermore, using templates and reference projects in DataOps.live ensures consistency and adherence to best practices. Template projects come pre-configured with standard directory structures, required files, and common configurations, reducing the setup time for new projects. Reference projects act as a guide for maintaining uniformity across multiple projects, as they can link standard configurations such as pipeline settings and stages. This approach ensures that teams can avoid reinventing the wheel and maintain high standardization, which is vital for scalability and maintenance.

Finally, DataOps.live's support for collaboration within projects enhances team productivity. With features such as role-based access control, project owners can assign specific permissions to different team members, ensuring everyone has the right access to the resources they need. This collaborative structure empowers teams to work together efficiently, whether it's data engineers, data scientists, or business analysts, ensuring that everyone contributes to the data operations lifecycle in an organized and secure manner.

By centralizing and structuring workflows within DataOps.live Projects, organizations can manage their data pipelines more effectively, reduce complexity, and foster better team collaboration. This approach improves operational efficiency and enables companies to scale their data operations in a controlled, governed manner, ultimately driving better decision-making and business outcomes.

Pipelines and Jobs

The Pipelines and Jobs section in DataOps.live is a foundational aspect of the platform, helping teams manage and automate the execution of complex workflows. A DataOps pipeline is a sequence of jobs configured to execute in a defined order, and these jobs can be managed, modified, and tracked through the platform. Pipelines are the driving force behind a project's data operations, orchestrating the execution of various tasks, from data extraction to transformation, modeling, and deployment.

A pipeline file in DataOps.live is a configuration file that details how the jobs should be run, including which stages they belong to and how they interact. Each pipeline file typically contains multiple jobs, which can be customized to suit the specific needs of the data operations workflow. This flexibility makes DataOps.live a powerful tool for teams looking to automate their data pipelines efficiently.

Jobs within a pipeline define individual tasks to be executed. These tasks may involve loading data, running transformations, testing, or any other task involved in data processing. They can be set up to run in parallel within a pipeline stage or sequentially across multiple stages. Configuring jobs with variables that control their behavior allows teams to adapt the execution dynamically, creating a more agile and responsive data workflow.

Each job can be configured to run in a specific environment, using a Docker container as an orchestrator. This container ensures that the necessary dependencies and tools are available for the job to execute correctly. The DataOps runner picks up jobs and executes them in the container, ensuring that everything runs as expected, regardless of where the pipeline is being executed.

Furthermore, the platform allows users to define base jobs, reusable templates for everyday tasks across multiple jobs. This reduces redundancy in job configurations and helps maintain consistency across different workflows. By using base jobs, teams can streamline their pipelines and focus on more specific customizations when needed.

The platform also supports creating secure, encrypted files called Vaults during each pipeline run. These Vaults store sensitive information, like API keys or database credentials, ensuring that they are protected while still being accessible to jobs as needed. Using Vaults adds an essential layer of security to DataOps.live, making it easier to manage secrets within the platform.

In DataOps.live, pipeline execution can be monitored, allowing teams to track job status and troubleshoot any issues. Visibility into each job's performance helps teams ensure their data pipelines function as expected, with minimal downtime or disruption. This aspect is critical for maintaining operational efficiency and providing insights into the health of data workflows.

Combining these elements makes Pipelines and Jobs a critical feature for teams using DataOps.live. This feature ensures that data operations are streamlined, scalable, and secure, allowing for detailed configuration, automation, security, and monitoring. Whether for simple or complex data workflows, DataOps.live's pipeline management capabilities are essential for modern data teams seeking to optimize their data processes.

Conclusion

The chapter provided an in-depth exploration of key concepts within the DataOps.live platform, emphasizing the essential elements for building and managing data projects. Understanding the roles of accounts, groups, subgroups, pipelines, and jobs within the platform lays the foundation for teams to efficiently structure and execute their data operations. By leveraging the powerful capabilities of DataOps.live, organizations can streamline their workflows, optimize project settings, and enhance collaboration among data teams.

As teams integrate these elements into their data practices, they can significantly improve speed and accuracy. DataOps.live allows for seamless project organization, automated pipeline execution, and a high degree of flexibility in managing complex data transformations. This ability to rapidly deploy and manage data workflows is critical in today's data-driven world.

One of DataOps.live's standout features is the ability to customize projects and adjust settings based on specific needs. By utilizing various components such as vault configurations, structured pipelines, and integrated job execution, teams can ensure their data processes remain agile, secure, and optimized. Whether managing small projects or scaling up for enterprise-level deployments, DataOps.live supports the adaptability required for modern data operations.

The focus on best practices for configuring and structuring projects also plays a pivotal role in ensuring long-term success. By following these guidelines, teams can avoid common pitfalls associated with disorganized data projects, ensuring their workflows remain efficient and scalable. As data grows in volume and complexity, DataOps.live equips organizations with the tools they need to stay at the forefront of data innovation.

CHAPTER 14

Getting Started with DataOps.live

Getting started with DataOps.live provides a robust foundation for integrating DataOps principles into data workflows. The platform simplifies data management by enabling efficient, automated, and scalable data operations, allowing teams to focus on value generation. Whether you are an experienced DataOps practitioner or a newcomer, this chapter will guide you through the initial steps to get DataOps.live up and running with Snowflake Marketplace and provide an overview of key features.

This chapter will walk you through obtaining DataOps.live from Snowflake Marketplace, setting up your environment, and ensuring seamless integration with your existing infrastructure. Following these steps will unlock the platform's full potential, enabling collaboration across teams, improving pipeline management, and fostering agile data operations from the get-go.

Acquiring DataOps.live from Snowflake Marketplace

To begin using DataOps.live, the first step is to acquire it from Snowflake Marketplace. The process involves navigating the Snowflake Marketplace interface and selecting the DataOps.live solution, allowing easy integration into your existing Snowflake setup. Once you've chosen DataOps.live, you can quickly subscribe to the service, streamlining the initial installation process. Snowflake's marketplace platform ensures secure, seamless, and reliable solution deployment into your environment.

R. L. Steelman, Jr., *Mastering Snowflake DataOps with DataOps.live*, https://doi.org/10.1007/979-8-8688-1754-0_14

The integration with Snowflake is vital because it enables organizations to leverage Snowflake's powerful cloud data platform capabilities while benefiting from DataOps.live's data operations management tools. By leveraging the native connection to Snowflake, DataOps.live can work efficiently with Snowflake's vast data warehouse ecosystem and orchestrate data pipelines, transformations, and tests directly within the Snowflake environment.

After subscribing to DataOps.live from the marketplace, you will have access to the necessary resources and documentation to configure and begin working with the platform properly. Snowflake Marketplace provides a secure and managed approach to software delivery, giving you the confidence to integrate DataOps.live into your infrastructure with minimal risk.

Setting Up Your DataOps.live Environment

Once you subscribe to DataOps.live, the next critical step is configuring your environment. This involves setting up accounts, groups, and projects within the platform, which is foundational to managing your data workflows. Start by ensuring that your Snowflake account is properly linked to DataOps.live. This integration allows DataOps.live to pull from Snowflake's databases and execute tasks such as transforming data and managing pipelines.

The process also includes configuring the DataOps Runner, a key component responsible for executing the data operations pipeline jobs. The DataOps Runner allows for the automation of tasks, ensuring that your data pipelines run smoothly. Additionally, you will configure your project settings, define your pipeline configurations, and ensure that your environment variables are properly set.

Within DataOps.live, it's also important to establish your CI/CD processes to allow for continuous integration and deployment of your data workflows. Setting up these processes will enable you to continuously update and monitor your data pipelines, which is essential for maintaining operational efficiency and ensuring that your data operations are always aligned with the latest business requirements.

Integrating DataOps.live into Your Data Workflow

With your DataOps.live environment configured, the next step is integrating it into your existing data workflows. DataOps.live is designed to work with various data orchestration tools and technologies, such as DBT and Snowflake, to streamline the data pipeline lifecycle. This integration allows you to manage your data transformations, testing, and deployment directly within the DataOps.live platform.

DataOps.live simplifies the orchestration of data pipelines by automating everyday tasks like data validation, data quality checks, and transformation jobs. Using built-in integrations with DBT, Snowflake, and other tools, you can create complex data workflows that ensure data integrity, improve team collaboration, and reduce manual intervention.

The platform also allows for version control and collaboration, enabling your teams to track changes to models, configurations, and pipelines over time. By leveraging these capabilities, you can ensure that your data workflows align with business needs and easily adapt to evolving requirements or new data sources.

Managing Data Pipelines and Jobs in DataOps.live

Managing data pipelines and jobs is a core component of DataOps.live and crucial for ensuring data operations run smoothly. Within the platform, pipelines automate data flow from one stage to another, while jobs define specific tasks within these pipelines, such as loading, transforming, or validating data. Integrating pipelines and jobs allows you to automate and monitor your data workflows efficiently.

DataOps.live makes it easy to configure and manage your pipelines, which can be customized to suit various business use cases. Whether you need to extract data from different sources, transform it, or load it into downstream systems, DataOps.live offers the flexibility to define and execute these tasks automatically. Additionally, job management features like scheduling, retries, and logging provide a layer of control and visibility, helping you identify issues and optimize your workflows.

By fully utilizing DataOps.live's pipeline and job management capabilities, you can scale your data operations to handle large datasets, complex transformations, and high-frequency updates. This functionality is key to ensuring that your data pipelines are not only efficient but also robust and fault-tolerant.

Scaling and Monitoring Your DataOps.live Projects

As your data operations grow, it becomes increasingly important to scale and monitor your projects effectively. DataOps.live is built to support enterprise-level deployments and can scale to meet the demands of large, complex data ecosystems. You can add new projects, pipelines, and jobs without worrying about compromising performance or reliability.

The platform provides various monitoring tools to track the health of your data pipelines and identify potential bottlenecks or failures. Dashboards and logs allow you to visualize pipeline execution and job status in real time, giving you a comprehensive view of your data operations. These monitoring tools are essential for maintaining high uptime, quickly identifying issues, and ensuring that your data workflows continue to run smoothly even as your projects scale.

Furthermore, DataOps.live supports the automation of data pipelines, allowing teams to work more efficiently while minimizing manual intervention. Automated monitoring and alerting ensure that teams are promptly notified of issues, enabling them to take corrective action quickly and avoid data disruptions.

Best Practices for Getting the Most Out of DataOps.live

To maximize DataOps.live's potential, adopting best practices throughout your data operations is essential. One key practice is continuously iterating on your pipelines using feedback from monitoring and testing. Regularly reviewing the performance and health of your data workflows ensures that you can quickly identify areas for improvement and optimize your processes.

Another best practice is to leverage version control and collaboration features within the platform to track changes, document decisions, and enable smooth collaboration across teams. This ensures that your data operations remain organized and all stakeholders are aligned on goals and processes.

Finally, ensuring that your security and compliance requirements are met when using DataOps.live is essential. By taking advantage of built-in security features, such as role-based access controls, audit logging, and secure integration with your data systems, you can maintain the integrity of your data while complying with industry standards.

Conclusion

Getting started with DataOps.live offers a comprehensive approach to managing data workflows and integrating DataOps principles into your organization. Following the steps outlined in this chapter, you will be well-equipped to deploy DataOps.live, manage your data pipelines, and scale your data operations. With a strong foundation, you can focus on optimizing your data workflows, improving collaboration across teams, and ensuring that your data operations align with business needs.

CHAPTER 15

Managing Your Environments

Effective environment management is one of the foundational elements in the successful application of DataOps principles. In the context of DataOps.live, environments are the isolated spaces where your data pipelines run, and they play a crucial role in maintaining clean workflows for development, testing, and production stages. Understanding how to manage and segregate these environments ensures seamless collaboration across teams while maintaining the integrity and performance of data operations. This chapter will delve into the intricacies of managing DataOps environments, exploring the different types of environments and their significance.

Managing your environments effectively ensures that each stage of the data lifecycle, from ingestion and transformation to testing and production deployment, operates without interference. The power of DataOps lies in the ability to automate workflows, trigger processes in the correct sequence, and keep environments consistent for different data projects. By using well-defined environments, teams can work in isolation, which reduces the risk of unintended changes impacting the entire system. This chapter will break down the importance of environment segmentation, how to properly configure environments, and the best practices for handling the complexities of multi-environment workflows.

Understanding the Key Types of Environments

In DataOps.live, the environment structure is carefully designed to separate data operations based on their purpose and lifecycle stage. There are four main types of environments: Production (PROD), Quality Assurance (QA), Development (DEV), and Feature Branches. Each environment plays a vital role in ensuring the data operations can scale without sacrificing quality or consistency.

R. L. Steelman, Jr., *Mastering Snowflake DataOps with DataOps.live*, https://doi.org/10.1007/979-8-8688-1754-0_15

The Production (PROD) environment is the most critical and permanent stage in DataOps.live. It houses the primary data pipelines, the production database, and other essential resources. The production environment is set up to ensure stable, high-quality data operations, running long-term processes essential for the organization's data-driven decision-making. Because the production environment is persistent, it is designed to minimize disruptions and create an ongoing, stable pipeline for real-time data.

The Quality Assurance (QA) environment mirrors the production environment, but it is used to verify that changes made during development and testing do not break the system. It serves as the last line of defense before new data processes are moved into production. The QA environment typically uses a similar setup as production but allows for testing and validation before deployment to minimize the risk of errors. In contrast, the Development (DEV) environment is much more transient and disposable, recreated every time a pipeline runs. It uses cloned data from the production environment to ensure that development happens on the most recent and accurate data without disrupting production systems.

The Feature Branches environment tests new features or performs experimental data operations. These branches are isolated and treated like miniature projects that contain their unique environment setup. By running pipelines on these feature branches, teams can validate changes without affecting the rest of the production or QA systems. These feature environments help teams accelerate innovation while controlling the primary data production processes.

Best Practices for Environment Segregation and Configuration

When configuring environments, it is essential to follow best practices to ensure that the transitions between environments happen seamlessly and without confusion. The environment segregation in DataOps.live ensures that each type of environment functions independently, and data from one environment doesn't impact others unless explicitly intended. This separation allows development teams to test and refine new data models or pipelines without worrying about side effects or unintentional consequences in production systems.

Best practices for segregating environments also include defining clear boundaries and naming conventions. For instance, different Snowflake accounts are assigned to each environment, and the naming of databases and schemas should follow a consistent naming convention to minimize confusion. Each environment should have its configuration settings and variables, ensuring that resources are optimized for their specific purpose. This isolation between environments prevents accidental data leaks and ensures the integrity of each stage in the data lifecycle.

Moreover, it is essential to ensure that pipelines are directed to the correct environment based on the branch from which they are triggered. By doing so, developers and data teams can ensure their changes are properly tested before moving to production. These carefully planned workflows also help manage the complexity of working with large-scale data operations, especially when multiple teams are involved.

Creating DataOps Environments

Creating DataOps environments in DataOps.live is fundamental in managing workflows and automating data pipelines effectively. By setting up environments, organizations can develop isolated spaces to manage different stages of data operations, such as development, testing, and production. Each environment in DataOps.live allows you to configure and control the resources and data processes necessary to ensure smooth operations. Environments serve as crucial components for ensuring that each stage of the data pipeline is efficiently managed, reducing risks in production and facilitating continuous development.

Creating DataOps environments in DataOps.live is straightforward and integrates tightly with Snowflake, ensuring that you can use familiar tools while benefiting from automation and best practices in data operations. When creating an environment, you'll begin by choosing the appropriate configuration for the specific task, such as connecting to your Snowflake account, defining the data storage settings, and customizing the environment's scope based on your organization's needs.

Each environment in DataOps.live can be tailored with specific access levels, configurations, and environment variables that align with the project requirements (Figure 15-1). This setup ensures teams can work on separate features or modifications without interfering with the production environment. Moreover, each environment can be linked to different Snowflake databases or schemas, making managing complex multi-layered data architecture in a unified platform easier.

pipelines/includes/default/snowflake_lifecycle_env.yml

```
Set Up Snowflake (PROD/QA):
extends: .agent_tag
image: $DATAOPS_SNOWFLAKEOBJECTLIFECYCLE_RUNNER_IMAGE
variables:
  LIFECYCLE_ACTION: AGGREGATE
  ARTIFACT_DIRECTORY: snowflake-artifacts
  CONFIGURATION_DIR: dataops/snowflake
resource_group: $CI_JOB_NAME
stage: Snowflake Setup
script: /dataops
environment:
  name: $CI_COMMIT_REF_NAME
  url: $SNOWSIGHT_URL/#/data/databases/$DATAOPS_DATABASE
artifacts:
  when: always
  paths:
    - $ARTIFACT_DIRECTORY
icon: ${SNOWFLAKEOBJECTLIFECYCLE_ICON}
rules:
  - if: '$CI_COMMIT_REF_NAME == "master" || $CI_COMMIT_REF_NAME == "main" || $CI_COMMIT_REF_NAME ==
"production" || $CI_COMMIT_REF_NAME == "prod" || $CI_COMMIT_REF_NAME == "qa"'
    when: on_success
  - when: never
```

Figure 15-1. *Sample Environment Configuration*

One key benefit of creating separate environments is the ability to define clear boundaries between different data pipeline stages. For example, in the development environment, data scientists and engineers can work on new features, write code, and test new transformations without impacting the integrity of the production data. Meanwhile, the production environment remains stable and ready to process real-time data, ensuring minimal disruptions to ongoing business operations.

Creating environments within DataOps.live reduces operational risks and provides complete visibility into how data flows across different stages. The platform offers monitoring and logging tools that track activity and provide insights into performance, enabling teams to identify bottlenecks or issues within each environment quickly. This transparency is critical for ensuring data operations remain efficient and reliable, especially when transitioning from development to production.

Creating DataOps environments also enhances collaboration among teams working on different data lifecycle stages. With multiple environments, developers can focus on building and testing new functionalities in isolation, while data analysts can work with clean, stable data in the production environment. This collaboration is streamlined, as DataOps.live integrates tightly with version control and allows team members to coordinate changes across environments seamlessly.

Another critical aspect of creating DataOps environments is maintaining consistency. By defining specific configurations for each environment, DataOps.live ensures that data processing, transformations, and load sequences follow the same standards across different stages. This standardization ensures that data remains accurate, relevant, and synchronized as it progresses through the various environments, reducing the chances of errors when moving data from one environment to the next.

Finally, creating environments in DataOps.live provides a scalable approach to managing data pipelines. As data needs grow and evolve, new environments can be added to accommodate expanding workflows. Whether for additional testing, new data sources, or increased team collaboration, the platform makes it easy to replicate existing environments or create entirely new ones. This scalability is crucial for businesses that adapt quickly to changing data demands while maintaining operational efficiency.

By carefully creating and managing environments in DataOps.live, organizations can ensure that their data pipelines remain organized, secure, and efficient across all stages, from development to production. This process sets the stage for successful data operations, improving data quality, accelerating time to market, and minimizing risks associated with complex data workflows.

Branching Strategies

Branching strategies are an essential component in DataOps, as they offer a structured method for managing the lifecycle of data projects. They enable teams to separate concerns, test features independently, and ensure stable deployments. A well-defined branching strategy ensures that teams can develop, test, and deploy features in a controlled and predictable manner. In DataOps, this process is crucial because of the dynamic nature of data pipelines, data transformations, and integration with various systems like Snowflake.

One of the most fundamental branches in DataOps workflows is the production branch. This branch serves as the stable version of the project and contains the final version of code and configurations that are deployed in the live environment. In a data-driven project, the production branch is where the live data pipelines are executed, and any changes that affect this branch are considered high-stakes and must undergo rigorous testing and approval processes. Ensuring that the production branch is always in a deployable state is crucial, as it directly impacts the users and operations relying on the data being processed.

Alongside the production branch, the quality assurance (QA) branch is equally vital. The QA branch acts as a staging environment for testing all new features and fixes before they are merged into production. It mirrors the production environment closely, providing a safe space to test the impact of new changes on the data pipelines, transformations, and integrations. It is important for teams working on data pipelines to test how new data flows behave in a near-identical setting to production before making them live. The QA branch ensures that no unforeseen issues could disrupt operations when the changes go live.

The development branch is another critical aspect of a branching strategy. It is where developers can work on new features, bug fixes, and other tasks that need to be developed independently. Changes in the development branch are usually not ready for deployment, and the branch is frequently updated as developers push new code. This environment allows developers to experiment and implement changes without fear of disrupting production or QA workflows. Moreover, the development branch is a buffer between feature development and the more structured testing environments, such as QA.

One of the most granular branches used in DataOps is the feature branch. This branch type is temporary and is used for developing specific features or changes in isolation from the main codebase. Feature branches are created for each task, whether adding a new pipeline functionality, implementing a data transformation, or working on a specific bug fix. Feature branches are typically short-lived and exist only for the task's duration. Once the work on a feature is complete, it is merged into the development branch, and the feature branch is deleted. This helps keep the codebase clean and free from unnecessary clutter.

In some workflows, especially for complex projects, cherry-picking is used to move specific changes between branches. For instance, if a particular feature developed in the development branch is ready for testing but others are not, cherry-picking allows developers to move just the relevant changes into the QA branch. This technique ensures that the development process remains flexible, allowing quicker testing of parts of a more significant task without waiting for the entire feature to be complete. Cherry-picking can also resolve conflicts between branches by manually selecting the changes that should be preserved.

Another significant advantage of a well-structured branching strategy is managing data environments that mimic the production environment without risking disruption to the live system. For example, when a feature branch is created, a corresponding replica of the data warehouse can be spun up to reflect the structure and data found in the production environment. This allows teams to conduct integration tests and data validations on real-world data, ensuring that new features or fixes work seamlessly in the production environment before deployment.

In a data-driven project, such as one utilizing Snowflake or similar platforms, branching strategies are further critical. Data pipelines and integrations can be complicated, requiring the alignment of many different components. For example, a transformation added in the development branch could have unforeseen consequences on downstream processes, such as aggregating data or moving data between databases. With proper branching strategies, these changes can be isolated and their effects tested before affecting the entire system. Furthermore, creating replicas of production environments for each branch ensures that the data stays consistent across development, testing, and production.

Branching strategies foster collaboration for teams working across different departments, such as development, data engineering, and analytics. Developers can work on their tasks independently without disrupting the work of others, and when their work is ready, it can be merged into shared branches for further testing. This isolation of work also means that data engineers can experiment with new transformations and data models, while analytics teams can analyze results without interfering with ongoing development. The result is smoother workflows and a more collaborative environment across teams.

Pull requests (PRs) and merge requests (MRs) play an important role in enforcing a disciplined and review-based approach to merging changes. When developers finish working on their feature branches, they initiate a pull request to merge the changes into the development or QA branches. Pull requests allow for thorough code review, ensuring that changes meet standards and do not introduce bugs into the system. For data projects, these reviews are not limited to code; they often include reviewing data models, transformation logic, and pipeline configurations.

DataOps teams also need to maintain an efficient process for conflict resolution. As multiple developers work on separate branches, there is a potential for conflicts, especially when changes are made to shared resources like data schemas or pipeline configurations. In such cases, it's essential to merge changes carefully and resolve conflicts manually, ensuring that the final codebase is stable and consistent. Conflict resolution practices are a crucial aspect of the branching strategy and help maintain the integrity of the overall project.

Lastly, proper management of branching strategies helps ensure that deployment to production happens in a controlled, predictable manner. Each step in the deployment process is well-documented, and the changes go through multiple layers of testing before being merged into production. This structured approach reduces the risk of introducing errors into the live environment, particularly for data-driven projects, where a small mistake can have large-scale implications. By clearly defining each branch's role and using automated processes to manage merges, DataOps teams can ensure a reliable and scalable deployment process.

Conclusion

Branching strategies are an integral component in the DataOps lifecycle, facilitating collaboration, testing, and deployment of data projects. By establishing clear and structured branching practices, teams can effectively manage the complexity of data workflows. With dedicated branches for production, QA, development, and feature-specific tasks, teams can ensure stability while promoting flexibility in development. The use of feature branches and other isolated environments allows teams to test in conditions that replicate production without the risk of disrupting ongoing work.

Managing multiple branches also enables parallel development efforts, where teams from different departments can work concurrently on different aspects of the data pipeline. With careful attention to pull requests and merge requests, developers and data engineers can review changes systematically and integrate them into the broader project with minimal risk. This approach reduces errors and ensures that only tested, validated code reaches the production stage.

Equally important is the practice of conflict resolution, which ensures that multiple changes are harmonized before they impact the live environment. DataOps strategies that include automated conflict management and detailed documentation further enhance the team's ability to deliver reliable results. The integration of automated tools ensures that potential issues can be detected early and addressed proactively, leading to a more efficient workflow.

With branching strategies in place, teams can also improve their ability to scale their workflows, enabling them to handle larger volumes of data, more complex data transformations, and faster-paced development cycles. The approach supports both smaller teams working on isolated changes and larger, cross-functional teams collaborating on multifaceted data initiatives. This scalability ensures that DataOps practices remain effective even as the complexity of the data architecture grows.

Additionally, clear branching strategies contribute to the creation of robust data pipelines that can be tested and deployed more efficiently. When combined with version control systems and integrated testing tools, teams can ensure that their data pipelines meet both functional and performance expectations. This is especially critical in data-driven environments, where even small discrepancies can have significant repercussions on the outcome.

By implementing these best practices in branching strategies, organizations ensure the successful deployment and ongoing maintenance of data pipelines. The synergy between development, testing, and production branches facilitates smoother operations and reduces the risk of costly errors. In turn, this structured approach leads to more reliable data systems that empower teams to make data-driven decisions with confidence, fostering an overall culture of trust and efficiency in data management.

CHAPTER 16

Build Your First DataOps Pipeline

Building your first DataOps pipeline is an exciting step toward automating and optimizing data workflows. This chapter introduces the foundational steps to construct a fully functional pipeline within the DataOps.live platform. By the end of this process, you will understand how to easily automate the extraction, transformation, and loading (ETL) of data, while ensuring that the pipeline is efficient, reliable, and scalable.

The chapter will guide you through the core concepts and tools essential for getting started, including pipeline creation, configuration, and managing tasks within the pipeline. You will gain insight into how the DataOps.live platform integrates with other systems and technologies, such as Snowflake, to streamline the process.

With a focus on practical implementation, the chapter aims to equip you with the knowledge needed to build pipelines that will automate your data operations. Whether you're new to DataOps or looking to expand your expertise, this chapter serves as a solid foundation for leveraging the power of DataOps.live in real-world scenarios.

Running Your First Pipeline

Running your first DataOps pipeline within the DataOps.live platform is a key step toward streamlining your data workflows and automating ETL processes. After setting up your project and configuring the necessary resources, the next step is to run your pipeline and verify its functionality. The process begins with ensuring that all dependencies and configuration files are in place and that your pipeline is properly connected to the data sources, targets, and any associated components like transformation logic.

R. L. Steelman, Jr., *Mastering Snowflake DataOps with DataOps.live*, https://doi.org/10.1007/979-8-8688-1754-0_16

To run your pipeline, navigate to the pipeline management interface within the DataOps.live platform (Figure 16-1). Here, you will find options to trigger the execution of your pipeline either manually or programmatically. This step ensures that data flows smoothly from source to destination, undergoing any necessary transformations. Each pipeline will be executed in stages, with progress being tracked in real-time, allowing you to monitor the success or failure of each task.

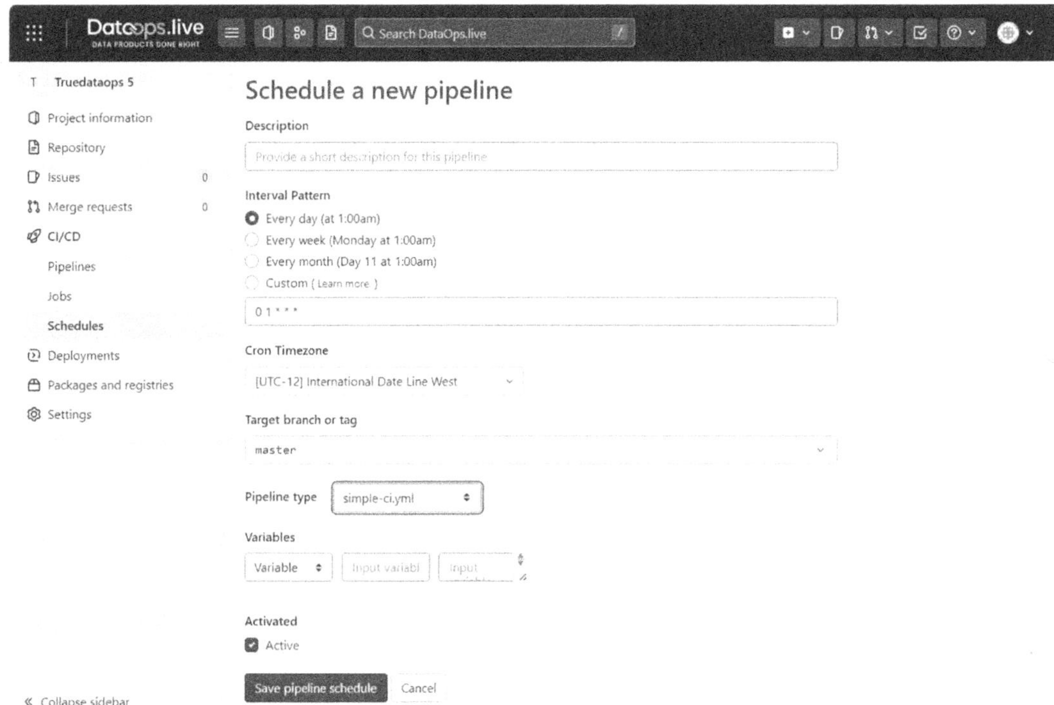

Figure 16-1. *Schedule a New Pipeline*

Once triggered, the platform handles orchestration automatically, executing each step as defined in your pipeline configuration. This includes extracting raw data, applying transformations, and loading it into the appropriate storage or database. Depending on your setup, this may involve leveraging technologies like Snowflake, where DataOps.live ensures compatibility for seamless integration. DataOps.live's robust logging and error-handling capabilities will capture any issues that arise during pipeline execution, enabling you to troubleshoot and fix errors quickly.

As the pipeline runs, you can view detailed logs and performance metrics, including execution times, resource usage, and any issues encountered. These logs provide essential insights into your pipeline's efficiency, helping you identify bottlenecks or other inefficiencies. If the pipeline encounters an error, the system will halt the execution and provide detailed feedback, allowing you to resolve the issue quickly.

When running your first pipeline, it's also essential to understand the process's lifecycle, from initiation to completion. Once the pipeline has successfully executed, DataOps.live provides tools to inspect the resulting data, validate outputs, and ensure everything has been processed as expected. This step is critical for data quality assurance, confirming that the transformations and data loads have been completed correctly.

Furthermore, running the pipeline for the first time allows you to tweak configurations for optimization. For example, you can adjust the parameters that control how often the pipeline runs or modify how errors are handled. You can fine-tune the execution environment, such as changing memory or compute resource allocations, to ensure your pipeline runs more efficiently during future executions.

Once your pipeline completes successfully, you can automate its execution as part of a continuous integration/continuous deployment (CI/CD) process. By automating the pipeline, you can ensure that your data workflows continue to run with minimal manual intervention. DataOps.live integrates seamlessly with version control systems like Git, enabling automation by tying pipeline triggers to commits or changes in your data sources or transformation logic.

It is also critical to ensure proper versioning is in place to track changes made to your pipeline over time; by utilizing the version control features of DataOps.live, you can roll back to previous pipeline versions or apply incremental changes with minimal disruption. This is particularly useful in a dynamic development environment where data needs to evolve continuously.

Monitoring your pipeline is an ongoing process, and DataOps.live makes it easy to track and monitor pipeline health over time. The platform provides dashboard views of historical execution data, making it possible to proactively identify trends, spot issues, and adjust the pipeline for long-term performance improvements. These insights also help you balance resource usage, execution times, and reliability.

In addition to real-time monitoring, DataOps.live allows you to configure alerting mechanisms that notify the relevant team members if something goes wrong during pipeline execution. These could include failure alerts, warnings for slow execution, or data validation failures. The flexibility of these notifications ensures that you're always in the loop regarding the health of your data pipelines.

Finally, as you continue to run and refine your pipeline, you'll develop best practices tailored to your organization's needs. By understanding how each step in the pipeline interacts with the data and adjusting as necessary, you can optimize performance, reduce errors, and ensure the pipeline is robust and scalable for future data operations.

Conclusion

Running your first pipeline on DataOps.live is an exciting and essential step in becoming familiar with the platform's capabilities. Following the outlined process, users can automate complex data workflows and ensure their pipelines run with minimal manual intervention. This step is crucial for building efficient and scalable data operations that contribute to better decision-making and streamlined workflows.

The platform's robust error-handling and logging capabilities provide valuable insights during pipeline execution, allowing users to identify and address issues quickly. With real-time monitoring and detailed logs, DataOps.live ensures that any problems encountered are thoroughly documented, making troubleshooting a smoother process.

Automation plays a central role in ensuring that pipelines are triggered seamlessly, which ties directly into continuous integration and deployment practices. DataOps.live's integration with version control systems like GIT allows teams to automate pipeline execution in sync with code or data changes, ensuring consistency and reducing human error.

As users continue to run their pipelines, fine-tuning configurations and optimizing resource allocation will become vital. DataOps.live offers tools to monitor pipeline health, track performance trends, and make data-driven adjustments to improve execution efficiency. This flexibility enables organizations to scale their operations while maintaining high data quality.

With proper version control and alerting mechanisms, teams can continuously refine and improve their data workflows. DataOps.live's monitoring features allow teams to address potential issues before they become critical, making it easier to maintain high uptime and reliability across data operations.

Ultimately, running your first DataOps pipeline is just the beginning. As teams become more familiar with the platform, they can expand their pipelines, refine their processes, and leverage advanced features to automate and optimize their data operations at scale. This foundational understanding sets the stage for more complex, high-performance data workflows.

CHAPTER 17

Getting Started with SOLE

This chapter introduces the Snowflake Object Lifecycle Engine (SOLE) and its capabilities in managing and optimizing data products. This framework is designed to simplify the management of Snowflake object configurations, allowing organizations to organize their data products more effectively. SOLE brings a new approach by organizing data objects by domains, offering a more modular, self-describing configuration structure. This shift helps improve accessibility and accuracy across large infrastructures, addressing challenges in object management.

With SOLE, users can break away from traditional hierarchical structures, opting for a more flexible, domain-specific organization that fits their operational needs. This allows for streamlined management of large-scale Snowflake objects while supporting the creation and maintenance of data products. By using YAML-based configuration files, SOLE enables users to define Snowflake objects in ways tailored to their business functions, such as finance or marketing, enhancing collaboration and scalability.

Introduction to SOLE and Data Products

The Snowflake Object Lifecycle Engine (SOLE) is designed to enhance the management of Snowflake objects in a more organized and scalable manner. Integrating SOLE into your workflow allows you to manage data products by treating them as individual domains rather than managing each object as a separate entity. This domain-based organization streamlines data handling, making it easier to scale and track the lifecycle of objects within Snowflake. SOLE helps address complexities by creating modular configurations that reflect the needs of various organizational departments or functions, such as finance, marketing, or operations.

In traditional approaches, Snowflake objects are often managed monolithically, creating challenges when scaling, maintaining, and evolving data models. SOLE solves these issues by allowing users to organize and track objects according to

R. L. Steelman, Jr., *Mastering Snowflake DataOps with DataOps.live*, https://doi.org/10.1007/979-8-8688-1754-0_17

specific business functions. This modularity ensures that each data product can be independently managed, making it easier to iterate on and update data models while maintaining a clear, organized structure.

The introduction of SOLE empowers teams to work more efficiently with large datasets. It abstracts away some of the complexity involved in managing Snowflake objects and provides a framework for users to define, update, and track data products over time. Each data product is treated as a collection of associated Snowflake objects, and these products can be described in terms of business domains. This organizational approach helps facilitate collaboration across departments, ensuring that teams can quickly locate, update, and manage data objects relevant to their specific needs.

With SOLE, configuration management becomes a key part of the data product lifecycle. YAML-based configuration files allow users to define the attributes of Snowflake objects in a declarative manner, meaning that objects are described in terms of their purpose and use within the organization. This method of defining objects encourages consistency, transparency, and repeatability, reducing the chances of misconfiguration or errors during deployment and operations.

When integrating SOLE into your data operations, teams can focus on building high-quality data products without worrying about managing each Snowflake object. Using SOLE, users can ensure that data products are automatically versioned and tracked as they evolve. This functionality is essential for maintaining high standards in data governance and ensuring compliance with internal and external regulations.

Another advantage of SOLE is its support for the continuous delivery and deployment of Snowflake objects. With SOLE, data teams can set up automated processes for managing the lifecycle of data products, reducing the manual effort required for deployments. This is especially beneficial for organizations with large, dynamic data environments that must maintain agility while ensuring consistency and reliability in their data pipelines.

By providing a structured approach to organizing and managing Snowflake objects, SOLE helps ensure that data products are well-documented, easy to maintain, and ready for collaboration. The engine's emphasis on modularity makes it suitable for complex data environments, where scaling and flexibility are essential considerations. The ability to define data products by domain enables teams to adapt to changing business needs quickly while maintaining data integrity and transparency.

Ultimately, SOLE provides a more efficient way of managing data products in Snowflake by aligning data infrastructure with business domains. As a result, organizations can optimize their data workflows, collaborate more effectively across departments, and clearly understand how their data is organized and utilized within their Snowflake environment. Integrating SOLE into your data operations strategy offers a strong foundation for developing scalable, high-quality data products.

Setting Up SOLE for Data Products

Setting up the Snowflake Object Lifecycle Engine (SOLE) for data products is a multi-step process designed to optimize data management within Snowflake environments. First, you'll need to ensure that your Snowflake account is configured correctly and that you have the appropriate privileges to create and manage objects. It's important to understand the underlying architecture of SOLE, which treats data products as distinct entities consisting of multiple Snowflake objects (such as tables, views, and schemas), allowing for better tracking and management of data through its lifecycle.

The next step is to define the data products within SOLE. This involves structuring the Snowflake objects to align with your business domains. Each data product typically corresponds to a specific business function, such as marketing, sales, or finance. By grouping related Snowflake objects into a data product, you establish a modular and scalable way to manage your data architecture. This domain-based approach helps organize data more efficiently and makes it easier to manage and govern data as it evolves.

After defining the data products, you will configure SOLE using YAML-based configuration files. These files allow you to specify the metadata and attributes associated with the data products. The configurations might include details such as which Snowflake objects are included in the product, their relationships, and their lifecycle stages. Setting up these configurations properly ensures that SOLE can track the data product through its entire lifecycle, from creation and updates to eventual deprecation or deletion. This automated tracking reduces human error and improves data governance across the organization.

Once your data products are defined and configured, the next step is to manage the lifecycle of each product. SOLE provides the tools to monitor the status of the data products, allowing you to automate the process of moving them through various lifecycle stages. These stages could include development, testing, and production, each with its own validation rules and checks. By automating this workflow, you can ensure that data products are consistently tested and validated before being promoted to the next stage, significantly reducing the risk of introducing errors into your production environment.

Another key aspect of setting up SOLE for data products is versioning. SOLE enables version control for Snowflake objects, meaning you can keep track of changes to your data products over time. A new version is created each time a modification is made to a data product, ensuring you can roll back to previous versions if necessary. This feature is particularly valuable in fast-paced data environments, where frequent changes to data models and pipelines occur. Versioning ensures that you always have a record of the evolution of your data products, which is crucial for debugging and auditing purposes.

As you set up SOLE for your data products, it's essential to integrate automated testing. SOLE supports the automation of tests to ensure that data products meet the necessary quality standards before being moved to production. These tests are part of the lifecycle stages, providing continuous validation. By automating these tests, you can significantly reduce the manual effort required for data validation and improve your data products' overall quality and reliability.

In addition to testing, robust data monitoring is crucial when managing SOLE for data products. SOLE provides monitoring capabilities to track the health and performance of your data products. You can monitor the status of your data products across different stages and catch issues before they affect downstream users or applications. By having real-time visibility into your data products' health, you can proactively address potential problems and minimize disruptions to your operations.

Collaboration across teams is also an important factor when setting up SOLE. As data products are typically aligned with different business domains, ensuring that the teams responsible for these domains are involved in the setup and maintenance of SOLE is critical. For example, the marketing team may need to define the structure of a marketing data product. In contrast, the IT team may be responsible for ensuring that the configuration files are correctly applied. Clear communication and collaboration across teams ensure that the setup is done correctly and that data products continue to meet the needs of the business.

Once the setup process is complete, you must regularly update and maintain the SOLE configurations. Over time, your business's needs may evolve, so your data products may need to be adjusted or enhanced. SOLE allows for easy configuration updates, ensuring data products remain aligned with business objectives. Whether adding new objects to a data product or changing their relationships, SOLE ensures these changes are tracked and validated appropriately.

Setting up SOLE for data products also involves defining access control policies to ensure only authorized users can modify or view specific data products. Access control is critical to data security, especially when dealing with sensitive information. By using SOLE, you can define access levels at the product, object, or lifecycle stage level, ensuring that each user has the appropriate level of access for their role.

Once everything is configured, the benefits of using SOLE for managing data products become clear. The modular approach to organizing data products makes scaling and adapting your Snowflake environment easier as your business grows. With automated lifecycle management, versioning, and testing, you reduce the risks associated with manual data management and ensure a more streamlined and efficient workflow. Additionally, SOLE provides better visibility and governance, which is essential for maintaining high-quality data and meeting compliance requirements.

Lastly, monitoring your data products' performance and usage regularly is important. By leveraging SOLE's monitoring features, you can continuously track the effectiveness of your data products, identifying any potential bottlenecks or issues before they become critical. This ongoing monitoring helps to maintain the reliability and performance of your Snowflake environment, ensuring that your data products deliver value to the business without interruptions.

Setting up SOLE for data products is essential for improving data management, governance, and collaboration within your Snowflake environment. By defining, tracking, and managing data products throughout their lifecycle, you can optimize your data workflows, enhance collaboration, and ensure that your data remains of the highest quality. SOLE's capabilities provide a comprehensive and scalable solution for organizations looking to streamline their data operations and scale effectively.

Modular and Self-Describing Object Configuration

In DataOps, modular and self-describing object configuration provides a flexible, scalable framework for managing data products and workflows. This approach allows teams to define Snowflake objects in a way that is both reusable and adaptable to changing business needs. By organizing objects into self-contained modules, each data product can evolve independently, ensuring that changes in one part of the system don't disrupt the workflow. This modularization helps maintain consistency and integrity throughout the data lifecycle, allowing for easier updates and improvements.

Each module in the system is self-describing, meaning it contains all the necessary metadata and configuration details to function independently. This allows anyone interacting with the data product to understand its purpose, structure, and dependencies quickly. It also enhances collaboration, as teams working on different modules don't need to rely on detailed documentation or coordination efforts to understand the objects they are working with. This level of transparency significantly reduces team friction and speeds up the development process.

The configuration's self-describing nature makes automation possible at scale. You can automate much of the data pipeline and lifecycle management by defining objects with clear metadata and dependencies. This reduces manual intervention and ensures data products are managed according to predefined rules. Automation improves the efficiency of data workflows and enhances data quality by minimizing human errors that could occur during manual configuration or updates.

Modular and self-describing configurations are also essential for version control. With this structure in place, each module can be independently versioned, ensuring that changes to one part of the system are tracked and managed without disrupting the entire data product. This versioning capability is vital in dynamic data environments where frequent changes to data models or pipelines are the norm. It also provides the ability to roll back to previous versions if a new change causes unforeseen issues, providing flexibility and stability to the data management process.

Furthermore, this configuration approach facilitates more straightforward integration with other tools and systems. Integrating Snowflake objects with external systems becomes more straightforward as they are modular and self-contained. Whether you're pulling data from an external source or pushing it to a different platform, the well-defined structure of the modules ensures that data can be accessed, updated, and shared efficiently. This level of interoperability is critical in modern data environments, where multiple systems often need to interact seamlessly.

Another significant advantage of this approach is its improved scalability. As data products grow in complexity or the amount of data increases, the modular design makes it easier to scale without affecting the entire system. New modules can be added or removed as needed, and existing ones can be updated without disrupting other system parts. This scalability ensures that the DataOps architecture remains agile and responsive to the evolving needs of the business.

Modular and self-describing object configurations also promote better data governance. With clearly defined objects and metadata, you can apply consistent governance policies across the system. Access controls, security measures, and data lineage can be integrated into the configuration, ensuring all data products comply with regulatory requirements and organizational standards. By managing data governance at the object level, you can enforce rules and policies that protect sensitive information while ensuring that data remains accessible and valuable to authorized users.

Additionally, this approach enables more efficient testing and validation. Each module can be individually tested, making it easier to identify issues and perform targeted testing rather than dealing with the complexity of an entire data product. This granularity in testing reduces the time and resources required for validation, allowing teams to focus on specific areas of concern without unnecessary overhead. It also makes it easier to ensure that changes to one part of the system don't negatively impact other areas.

In summary, the modular and self-describing object configuration in DataOps provides a robust, scalable, and efficient way to manage Snowflake data products and workflows. Organizations can streamline their data operations, enhance collaboration, and improve overall data quality by breaking complex systems into manageable, independent modules. This approach simplifies configuration and management and facilitates automation, version control, integration, scalability, governance, and testing, making it an essential practice for modern data environments.

Best Practices for Organizing SOLE Configurations

Organizing SOLE configurations effectively is essential for maximizing the efficiency and scalability of data operations. The first best practice is ensuring a clear separation of concerns in your configuration structure. This means organizing objects into distinct modules based on their function, such as ingestion, transformation, and validation; each module has a specific role within the data pipeline. By keeping configurations modular and focused, teams can work more efficiently and make changes without impacting other areas of the system.

Another best practice is to use standardized naming conventions. A consistent naming structure makes it easier for all team members to understand different objects' and configurations' purpose and context. This standardization helps in documentation, troubleshooting, and onboarding new team members. It also supports version control, as naming conventions can include version numbers or timestamps, providing a clear history of changes.

Incorporating automation into your configuration management strategy is also crucial. By automating repetitive tasks such as environment setup or data validation, teams can reduce manual errors and ensure consistent deployments across different environments. Automated testing and continuous integration (CI) processes should be integrated into the configuration to guarantee that changes are validated before deployment. This improves the reliability and efficiency of the entire system.

Version control is another critical best practice. Robust versioning strategies must be implemented for the configurations and data objects. This allows teams to track changes over time, roll back to previous versions when necessary, and manage multiple versions in parallel for different environments or use cases. Using tools like Git for version control in conjunction with the SOLE configuration can help maintain consistency and prevent configuration drift.

It's also essential to define clear access control policies. Not all team members should have the same level of access to the configurations or data products. By enforcing role-based access controls (RBAC), you can ensure that only authorized users can modify critical configurations or deploy changes. This reduces the risk of accidental errors or security vulnerabilities and helps maintain the integrity of the data products.

Documentation plays a key role in organizing SOLE configurations. Every module, object, and setting should be well-documented to provide clarity for all team members. Clear documentation ensures everyone understands the configuration's structure, purpose, and interaction with other elements. It also simplifies troubleshooting and makes onboarding new team members or external partners who need to understand the system easier.

Regular reviews and audits of the configuration are essential for maintaining long-term effectiveness. These reviews help identify areas for improvement, ensure compliance with internal standards or external regulations, and keep the configurations aligned with the business objectives. Conducting these audits periodically ensures that configurations remain optimized and relevant to changing business requirements.

Finally, always ensure that your configurations are adaptable to future needs. As your data architecture and business needs evolve, your SOLE configurations should be flexible enough to accommodate new data sources, transformations, or processes without requiring major overhauls. Building flexibility into the configuration from the outset can save significant time and resources as your organization grows and evolves.

Advanced Configuration Options and Use Cases

Advanced configuration options in SOLE (Snowflake Object Lifecycle Engine) offer several powerful features that enable organizations to customize and optimize their data operations. These configurations are designed to meet complex business requirements, streamline workflows, and ensure data products function optimally across different environments. Understanding these advanced options is crucial for teams looking to fully exploit SOLE's capabilities.

One of the advanced configuration features is the ability to define data object lifecycle management policies. These policies allow users to automate the process of moving data through its lifecycle stages, from ingestion to archival or deletion. Teams can specify rules for when data should be archived, when it needs to be cleaned, or when older versions should be purged, helping ensure that data storage costs remain controlled while maintaining regulatory compliance.

Another important configuration option is the ability to integrate custom data validation logic into the SOLE pipeline. This is particularly useful for teams working with large and complex datasets that require unique validation rules before they can be processed or shared. By defining custom validation functions, teams can automate the error-checking process and ensure that only clean and valid data flows through the system. This reduces the risk of introducing data quality issues into downstream processes.

Advanced configurations also support the integration of third-party data sources and tools. With SOLE, teams can configure data pipelines to interact seamlessly with external systems, whether cloud storage providers, data warehouses, or APIs. For instance, businesses that need to import data from external partners can configure their pipelines to fetch data at specified intervals or on demand, making data integration more efficient. Additionally, this functionality is critical for organizations that operate in multi-cloud environments.

Another notable feature of SOLE is the ability to configure and manage multiple environments within the platform. This functionality allows users to create distinct development, testing, staging, and production environments. Each environment can have its own set of configurations and data access policies, which makes it easier to deploy updates and test new features without disrupting the production environment. This approach fosters better collaboration among development teams and reduces the chances of introducing bugs into live systems.

In addition to environment management, SOLE offers flexible scheduling options for automated pipeline execution. Users can define complex scheduling patterns to determine when specific data operations should occur. This includes setting up triggers based on time, data availability, or external events. For example, a pipeline can be configured to run whenever a new dataset is uploaded to the system or after completing a previous task, ensuring that all processes are synchronized and optimized for performance.

For organizations with strict security and compliance requirements, SOLE provides advanced configuration options for managing access control and auditing. Teams can define granular access permissions for different roles, ensuring that sensitive data is only accessible to authorized personnel. Additionally, SOLE enables comprehensive audit logs, which track all actions performed within the system, helping organizations maintain transparency and comply with industry regulations like GDPR or HIPAA.

The SOLE platform also allows teams to implement custom notifications and alerts based on predefined conditions. This feature is invaluable for proactive monitoring of data pipelines. For instance, users can configure the system to send an alert whenever a pipeline fails, a threshold is exceeded, or when data quality issues are detected. These notifications help teams respond quickly to problems, reducing downtime and minimizing the impact of issues on data-driven processes.

Customizable error handling is another key feature of SOLE's advanced configuration options. Users can define how the system should respond to different types of errors. Whether retrying a failed task, sending an alert to an administrator, or skipping problematic records, this level of customization allows teams to design resilient workflows. By incorporating intelligent error handling, teams can ensure that data operations continue smoothly, even during unexpected failures.

SOLE's advanced configuration options for parallel processing and resource allocation can significantly improve pipeline performance for teams working with large-scale data sets. Users can configure pipelines to run multiple tasks in parallel, optimizing the use of computing resources and reducing the time required for data processing. This feature is especially beneficial when dealing with resource-intensive tasks like data transformations or complex machine learning workflows.

One of the standout features of SOLE's advanced configuration is its ability to integrate machine learning models into data workflows. Teams can configure data pipelines to include steps that invoke machine learning models for predictive analytics or anomaly detection. This enables businesses to automatically incorporate AI-driven insights into their data processing pipeline, helping to unlock value from their data in real time.

Lastly, SOLE offers extensive logging and monitoring capabilities, providing teams with real-time insights into pipeline performance. Advanced configuration options allow users to track metrics like job completion times, success rates, and resource usage. This data is invaluable for optimizing pipelines, identifying bottlenecks, and ensuring the system runs as efficiently as possible.

By leveraging these advanced configuration options, organizations can optimize their data workflows, improve performance, and scale operations effectively. Whether it's automating lifecycle management, integrating third-party tools, or incorporating machine learning models, SOLE provides the flexibility and power that data teams need to drive innovation and meet business demands.

Conclusion

The advanced configuration options within SOLE allow teams to design and manage complex data pipelines tailored to their needs. Organizations can ensure that their data operations are efficient and secure by integrating robust lifecycle management, custom validations, and access controls. These features allow for better resource utilization, quicker response times to data anomalies, and increased scalability in the face of growing data needs.

Automating tasks, integrating third-party tools, and applying granular access controls improve overall pipeline management. By leveraging SOLE's scheduling, monitoring, and error-handling capabilities, teams can minimize downtime and enhance their data operations' efficiency. The comprehensive auditing and logging features further ensure transparency and compliance, providing teams with the necessary tools to meet industry regulations.

Moreover, SOLE's integration with machine learning models allows businesses to make smarter, real-time decisions based on predictive analytics and anomaly detection. This enables teams to incorporate cutting-edge technologies into their workflows without compromising performance or scalability. As organizations continue to expand their data operations, the advanced configuration features of SOLE will provide the infrastructure needed to support their growth and innovation.

Ultimately, the flexibility and depth of SOLE's advanced configuration options allow data teams to tackle complex workflows, improve operational efficiency, and enhance the quality of their data products. By understanding and implementing these configurations, teams can unlock the full potential of their data operations, ensuring they remain agile and responsive to ever-evolving business needs.

CHAPTER 18

Getting Started with MATE

The Modelling and Transformation Engine (MATE) is integral to the DataOps.live platform, designed to streamline and automate the transformation of raw data into valuable insights. It is a versatile tool that empowers users to easily manage data transformation workflows, focusing on scalability and efficiency. MATE allows for seamless integration with various data sources, making it a critical component for organizations looking to enhance their data management processes.

With MATE, users can leverage pre-built transformations or create custom scripts to handle complex data transformations. This flexibility helps to ensure that the engine can be tailored to the specific needs of different industries or use cases. Whether working with small datasets or large-scale enterprise systems, MATE's robust capabilities ensure consistent performance and reliability.

The engine's design is centered around best practices in data transformation, including automation, version control, and orchestration. By integrating MATE into your data operations, teams can reduce manual interventions and accelerate time-to-insight while ensuring data integrity and governance standards are met. It supports various data transformation scenarios, including ELT, batch processing, and real-time streaming.

As part of the DataOps.live ecosystem, MATE complements the platform's data pipeline capabilities by providing an intuitive, scalable approach to data modeling and transformation. In this chapter, we will explore MATE's core features and configuration options and how it can be used to drive efficient and automated data workflows that align with DataOps principles.

MATE Versus DBT

MATE (Modelling and Transformation Engine) is distinct from DBT (Data Build Tool) in several ways, although both are designed to manage data transformations. MATE is built as part of the DataOps.live platform, designed to enhance the data transformation

R. L. Steelman, Jr., *Mastering Snowflake DataOps with DataOps.live*, https://doi.org/10.1007/979-8-8688-1754-0_18

process within a broader DataOps framework. Unlike DBT, which focuses primarily on transforming data within the data warehouse, MATE offers tighter integration with the entire DataOps pipeline, enabling more data processing and orchestration flexibility across various environments.

One significant difference is how MATE integrates into the DataOps.live platform. It emphasizes automation and real-time data transformation capabilities, providing a more comprehensive solution for managing data workflows end-to-end. This integration ensures that data processing, from extraction to transformation, is seamlessly orchestrated within a unified pipeline, a less emphasized feature in DBT.

MATE also provides built-in support for version control and automated testing, which can be more complicated to implement manually in DBT. This native integration allows teams to maintain robust governance, track changes, and manage real-time collaboration within the platform's ecosystem. Furthermore, MATE offers custom transformation scripts and scalable processing that allow for more advanced use cases than DBT's templated, SQL-focused approach.

Another key differentiator is MATE's handling of complex data models. While DBT strictly follows an SQL-based approach to transformation, MATE extends this by offering support for various languages and transformation frameworks. This flexibility allows teams to utilize different tools within their workflows based on specific project needs, all while maintaining a consistent and efficient process.

Ultimately, MATE is built for flexibility, scale, and seamless integration with DataOps principles. While DBT is excellent for SQL-based transformations, MATE's deeper integration within DataOps.live provides a more holistic approach to managing data transformations that supports larger, more complex data environments with automated testing, governance, and enhanced collaboration capabilities.

Core Concepts of MATE

The core concepts of MATE (Modelling and Transformation Engine) revolve around providing an efficient framework for transforming raw data into structured, actionable insights. MATE allows data engineers to define and manage data transformations robustly and flexibly. The foundation of MATE is the ability to model transformations using SQL and Python, facilitating adaptability across different use cases and technical environments.

Key elements of MATE include projects, sources, models, and materializations. A MATE project encapsulates all the necessary configurations for the data transformation process, defined primarily in a dbt_project.yml file. Sources describe the initial data sources from which models are built. Models represent the logic applied to source data through SQL or Python. Materializations control how data is stored and updated, allowing for flexibility in the system's management of transformed data.

YAML and SQL templates enable scalability and traceability throughout the entire transformation pipeline. Additionally, integrating Python models in MATE offers advanced capabilities, such as using external libraries for more complex transformations or frameworks like PySpark for distributed processing. This flexibility allows organizations to scale their data transformation efforts as needed while maintaining clear lineage and testing through defined sources and models.

In addition to foundational concepts, MATE supports the management of incremental data loads and the materialization of views and tables, allowing users to fine-tune their data models to meet specific performance and operational requirements. The incremental model configuration is particularly valuable for handling large datasets, ensuring that only updated records are processed in each pipeline run, saving processing time and resources.

MATE's design is based on the principles of modularity and reusability. It allows users to define models and sources modularly, encouraging best practices for data modeling. This modularity ensures teams can collaborate more effectively, leveraging shared models and sources without duplicating efforts.

Overall, MATE streamlines data transformation and modeling, providing a structured approach that seamlessly integrates modern data engineering workflows. Aligning with DataOps principles, it helps organizations automate and optimize data transformation, ensuring that data remains accurate, consistent, and accessible for analytics and reporting. Through its robust set of tools, MATE enhances the efficiency and scalability of data pipelines.

Building Models with MATE

Building models with MATE (Modeling and Transformation Engine) allows teams to define and transform data through an automated, modular system. One of MATE's key features is its ability to simplify complex data transformations, offering an intuitive interface to structure, organize, and manage models effectively. The process is designed for scalability and flexibility, enabling data teams to quickly adapt their models in response to business changes or new data sources.

MATE integrates seamlessly with the DBT framework, leveraging its transformations while enhancing workflow automation. Teams can define data models in code, which promotes collaboration and enables version control. This integration facilitates data-driven development by encouraging reusable, well-structured models. MATE's ability to work with different data sources and formats makes it an invaluable tool for building enterprise-level data solutions.

A unique feature of MATE is its support for dynamic modeling. Instead of relying on static definitions, models can automatically update based on changes in the data schema or underlying structure. This adaptability ensures that the models remain relevant and accurate, regardless of changes in source data or business requirements. As a result, organizations benefit from enhanced data consistency and accuracy.

The core principle behind MATE is its focus on automation, helping organizations reduce the manual overhead associated with data modeling and transformation. By automating many tasks, MATE frees up valuable time for data engineers and analysts to focus on higher-level tasks, such as analysis and strategic decision-making. The tool's ease of use and automation improve both efficiency and productivity when building data models.

MATE data models are built using a modular approach, meaning components can be reused across different models or projects. This modularity helps streamline development by minimizing redundancy and reducing the risk of errors. The result is a cleaner, more efficient data pipeline, where changes in one module automatically propagate through dependent models.

MATE's dynamic approach also includes self-describing models, where the model configuration is embedded directly in the code. This approach reduces the need for external documentation, as the model is self-contained, making it easier for developers to understand and manage. Furthermore, this eliminates the risk of configuration drift, as the system is always aligned with the codebase.

The transformation engine within MATE is optimized to handle large datasets and complex transformations efficiently. Parallel processing and optimized query execution ensure that even large-scale data models can be built and transformed rapidly. This results in quicker project turnarounds and allows organizations to easily handle big data challenges.

One significant advantage of MATE is its ability to manage a model's full lifecycle, from creation to deployment. It simplifies model testing, validation, and deployment, ensuring that models are fully functional and optimized before they are moved into production. MATE also provides mechanisms for tracking model changes so teams can maintain visibility into the model's evolution over time.

Teams using MATE also benefit from its integration with monitoring and alerting features. This allows them to track the performance of their models in real time, ensuring that any issues are identified and addressed quickly. The monitoring capabilities also help track data quality and consistency, which is vital for maintaining the integrity of data products.

Another essential aspect of MATE is its ability to generate documentation automatically. When building data models, MATE can produce documentation that includes schema definitions, transformations, and other critical information. This feature helps ensure that the knowledge within the models is accessible to the broader team, even those not directly involved in development.

In summary, MATE's combination of modularity, automation, dynamic adaptation, and integration with existing tools like DBT makes it a robust framework for building and managing data models. Its flexibility and ease of use help teams create more efficient data pipelines, reducing complexity while improving the data quality. As organizations continue to scale their data operations, tools like MATE will be essential in driving productivity and maintaining high data integrity and governance standards.

Using DBT Packages with MATE

To integrate DBT packages with MATE, users can leverage DBT's powerful transformation capabilities while benefiting from MATE's flexibility and automation features. DBT (Data Build Tool) is widely used to transform raw data into a format suitable for analysis. When combined with MATE, users can streamline the entire process of modeling, transformation, and deployment.

DBT packages are collections of pre-built models, macros, and tests that help users implement commonly needed functionality quickly. By incorporating these packages into MATE, teams can save time and effort in building custom transformations from scratch. MATE facilitates smooth integration by automatically including these DBT packages within data pipelines, reducing the complexity of data workflows.

One key advantage of using DBT packages within MATE is the ability to reuse industry-standard transformations. These pre-built packages ensure that best practices are adhered to, and users can focus on specific business logic rather than reinventing the wheel. MATE enhances this by enabling seamless management and deployment of DBT packages, streamlining the deployment process, and making it easier for teams to stay agile.

MATE's configuration system plays an essential role in integrating DBT packages. Users can easily add DBT packages to their data models without modifying the underlying infrastructure through its modular and self-describing configuration approach. MATE automatically resolves dependencies and ensures that each package is configured correctly, eliminating manual intervention and ensuring the transformation pipeline runs smoothly.

Moreover, MATE ensures that the integration of DBT packages is efficient and flexible. MATE allows users to configure their projects and environments to optimize the execution of DBT transformations. This flexibility ensures that teams can integrate DBT packages into various environments, such as development, staging, and production, while maintaining the necessary configurations across all environments.

Incorporating DBT packages within MATE also enhances model version control and collaboration. Since MATE integrates well with version control systems like Git, teams can track changes to their DBT packages, ensuring that transformations are reproducible and traceable. This is particularly valuable in environments where multiple data engineers and analysts work on the same project, as it helps prevent conflicts and improves collaboration.

MATE also extends the power of DBT packages by making them more accessible through its automated documentation features. When DBT packages are used within MATE, the system generates self-describing documentation that explains how each package interacts with the data models. This documentation provides transparency and makes it easier for new team members to understand the data transformation logic.

DBT packages benefit from enhanced monitoring and alerting capabilities when used within MATE. MATE's real-time monitoring tools allow teams to track the performance and health of their DBT transformations. This ensures that any issues with data transformations are immediately flagged, enabling quick resolutions and minimizing downtime. Alerts can be configured based on specific thresholds, such as execution time or error rates, to keep teams informed.

Additionally, MATE's integration with DBT packages allows for more efficient testing of data models. MATE's built-in testing frameworks can be used to validate DBT transformations and ensure data quality throughout the process. Automated testing reduces the risk of errors and inconsistencies, helping to maintain high-quality, accurate data.

Using DBT packages within MATE also accelerates the data pipeline deployment process. As DBT packages are pre-built and tested, they can be deployed quickly and without additional configuration. MATE automates the deployment process, reducing manual intervention and allowing teams to focus on business logic and innovation instead of infrastructure management.

Regarding scalability, MATE's handling of DBT packages supports large-scale data operations. MATE is designed to scale with the growth of your data infrastructure, making it easy to add more DBT packages as needed. As your data pipeline grows, MATE can efficiently manage an increasing number of transformations and integrations, ensuring that performance remains consistent even as workloads expand.

Finally, the synergy between MATE and DBT packages improves the overall efficiency and flexibility of the data team. Data teams can build robust, scalable data pipelines faster by combining the power of DBT's pre-built transformations with MATE's flexible, modular architecture. This integration allows teams to focus on delivering value through insights and business intelligence, rather than dealing with the complexities of manual data transformation and configuration.

Macros with MATE

In MATE, macros are reusable code blocks that simplify data transformations across multiple models. These code snippets, written in SQL or Jinja, help automate repetitive tasks, boosting efficiency and reducing redundancy. By centralizing commonly used logic, macros ensure consistency and ease of maintenance within data pipelines. For example, a macro could be created to standardize the formatting of dates, ensuring uniformity across different datasets.

Macros can also enhance modularity in your data models. Rather than repeating the same transformation logic in multiple places, you define it once in a macro and reuse it wherever necessary. For instance, a macro that calculates customer lifetime value can be applied across several models that need this metric, reducing errors and promoting collaboration among team members.

A key benefit of macros in MATE is their ability to streamline performance. By encapsulating complex logic into a single reusable function, you avoid duplication of effort and make optimizations in one place that benefit all models using the macro. For example, a macro designed for an aggregate operation can be fine-tuned, improving query performance across multiple data models without needing individual updates.

Another advantage is that macros in MATE allow for flexible, dynamic transformation logic. They can be written to handle more complex scenarios involving loops, conditionals, or custom functions. For instance, a macro could be created to calculate a running total based on varying parameters, such as the number of days or specific regions, providing flexibility in executing transformations.

Another essential feature is the ease of testing and debugging macros. When a macro is defined, it can be independently tested to ensure its logic works correctly before being applied in production models. For example, a macro for filtering data by certain conditions can be tested on small sample datasets to verify its functionality, making debugging simpler and less error-prone.

Lastly, macros in MATE improve overall data pipeline maintainability. By abstracting frequently used logic into a central location, updates and modifications become easier to manage. If a change needs to be made, it can be done in the macro, and the updates automatically reflect in all models that utilize it, reducing maintenance overhead and keeping the data pipeline efficient.

MATE Best Practices

To effectively leverage MATE, it is important to follow best practices that ensure maintainability, scalability, and efficiency within your data models. One key practice is modularization. Breaking down complex transformations into smaller, reusable components promotes easier maintenance and enhances readability. Instead of crafting a massive model that handles multiple tasks, splitting the work into distinct, manageable sections is better, allowing for flexibility and reducing the risk of errors.

Another important consideration is maintaining clear documentation. This helps team members understand the logic and transformations applied to each data model. Documenting macros, models, and pipelines makes it easier to troubleshoot, onboard new members, and ensure everyone is on the same page when changes are made. This practice can benefit teams working with large datasets or complex data environments, where understanding data flow is crucial.

Version control is also an essential practice when using MATE. As data models evolve, so should your approach to tracking changes. Using version control systems like Git ensures you can track who made changes, when, and why. This is vital for collaboration and for rollback purposes if something goes wrong. Version control allows you to keep historical snapshots of the model, which is beneficial when reverting to previous configurations or troubleshooting issues that might arise after changes.

Designing efficient models is crucial to avoiding performance bottlenecks. In MATE, you can optimize queries by using proper indexing and partitioning strategies. Rather than loading all data at once, break it down into smaller chunks and only pull the necessary datasets for the task. Efficient data handling ensures your transformations run smoothly, improving overall pipeline performance and reducing execution times. Using MATE's built-in tools to monitor and profile model performance can help pinpoint areas for optimization.

Testing is another best practice to follow when working with MATE. Implementing a robust testing strategy that includes unit tests for individual components and integration tests for larger models is vital. This ensures that each transformation and macro behaves as expected, avoiding issues when the model is pushed to production. Automated tests help catch problems early and provide confidence that the models work correctly before they are deployed at scale.

To streamline collaboration, it's important to establish a common framework and standard operating procedures (SOPs) for MATE use. Agreeing on naming conventions, model structures, and transformation patterns early on helps ensure consistency across the team. It reduces the chances of conflicts or confusion and enables team members to pick up where others left off. Additionally, leveraging MATE's features for team collaboration, such as project sharing and permissions management, ensures that everyone has the necessary access to contribute effectively.

Another practice is to embrace a data-centric approach to design. Instead of focusing solely on the technical implementation, prioritize the outcomes the data models aim to achieve. By considering how the end users interact with the data, you can design models that deliver value while avoiding unnecessary complexity. This mindset shift helps prevent over-engineering, ensuring that data models remain clear and focused on the business objectives.

Lastly, continuous improvement should be an ongoing practice. Regularly review and refine your models, macros, and data pipelines based on feedback and performance metrics. As data environments evolve, the models that worked initially may need adjustments to stay relevant. By constantly iterating and learning from each deployment, you ensure that the MATE models remain agile, adaptable, and aligned with business needs.

Conclusion

Understanding and implementing best practices is essential for achieving successful, scalable data models as you proceed with MATE. By emphasizing modularization, documentation, version control, and performance optimization, teams can streamline their workflows, mitigate potential risks, and ensure data models perform efficiently. Consistency in design and continuous testing further strengthens the overall quality and reliability of the data models.

Collaboration also plays a key role in ensuring that MATE is leveraged effectively. Clear communication, shared standards, and a unified approach within teams help to avoid confusion and foster an environment of innovation. Furthermore, integrating feedback loops and maintaining an iterative improvement process ensures that the models evolve in alignment with business needs.

Another important aspect of MATE is its ability to integrate well with existing tools and systems. Teams can build a robust and cohesive data pipeline infrastructure by combining MATE's flexibility with other technologies, such as DBT and Snowflake. This synergy enables teams to adapt quickly to changing requirements and deliver high-quality data products.

The ability to scale and adapt to evolving data needs is crucial in today's fast-paced, data-driven world. By following best practices and continuously refining your models, you ensure that MATE will meet your current needs and be capable of handling future challenges. This forward-thinking approach will set the foundation for long-term success.

In the end, implementing MATE effectively requires a combination of technical expertise, strong collaboration, and an understanding of the broader goals of the data ecosystem. By following the best practices outlined, you can ensure that your data models are efficient and sustainable, driving value across your organization while maintaining a high performance standard.

CHAPTER 19

Managing Multiple Databases with DataOps.live

Managing multiple databases is critical to modern data management, especially in complex environments where different systems or use cases require distinct databases. Within DataOps.live, the ability to efficiently manage multiple databases ensures that teams can handle diverse data sources, integrate them seamlessly, and leverage the strengths of each database system. This chapter will explore strategies for setting up and optimizing workflows across multiple databases in DataOps.live, enhancing performance, and maintaining consistency.

One key challenge in managing multiple databases is ensuring data flows efficiently between systems without introducing unnecessary latency or complexity. DataOps.live provides tools and methodologies to streamline this process, enabling teams to manage cross-database workflows seamlessly. The chapter will also cover best practices for maintaining data integrity, automating processes, and troubleshooting common issues that arise when working with multiple databases.

Throughout the chapter, we'll focus on practical steps for implementing and managing multiple databases in DataOps.live. Whether you're integrating databases for scalability, performance, or specialized use cases, this guide will provide valuable insights into optimizing cross-database workflows and maintaining best practices. By the end of this chapter, you'll know how to manage multiple databases confidently and ensure smooth data operations in a DataOps environment.

R. L. Steelman, Jr., *Mastering Snowflake DataOps with DataOps.live*, https://doi.org/10.1007/979-8-8688-1754-0_19

Understanding Database Management in DataOps.live

Database management in DataOps.live revolves around efficiently handling and integrating multiple data sources to ensure seamless workflows and accurate data pipelines. By supporting various database types, DataOps.live allows users to maintain flexibility when working with structured and unstructured data. The platform offers centralized database management, enabling teams to orchestrate complex data flows across different environments while ensuring data integrity and security.

The system's ability to manage databases across environments provides a significant advantage in scaling operations and adapting to business needs. Users can configure their workflows to interact with multiple data sources, ensuring data consistency across all systems. This is particularly important in environments where real-time data processing or near-instantaneous updates are critical to decision-making and operational efficiency.

Furthermore, DataOps.live facilitates the automation of processes that typically require manual intervention when managing multiple databases. Integrating pipelines across different database systems enables efficient data synchronization, reducing the risk of errors and ensuring that the data stays consistent across various platforms. This streamlines data operations and speeds up deploying changes across multiple systems.

Another essential aspect of database management in DataOps.live is the ability to troubleshoot and monitor database health. The platform provides users with tools to quickly diagnose data inconsistencies, connectivity problems, or performance bottlenecks. With these tools, teams can address issues proactively, preventing disruptions that may impact data workflows or end-user experiences.

DataOps.live also supports database version control, crucial in environments where databases evolve frequently. This feature allows users to track changes, roll back to previous versions when needed, and maintain a reliable history of database updates. It is particularly beneficial for maintaining data governance and ensuring all database changes align with compliance and security standards.

Organizations can efficiently manage and optimize their databases in DataOps.Live by understanding and leveraging these features. This approach leads to more reliable data workflows, improved operational efficiency, and the ability to scale operations without sacrificing data quality or performance.

Setting Up Multiple Databases in DataOps.live

Setting up multiple databases in DataOps involves configuring and managing several environments to support diverse data workflows. This process enables teams to orchestrate data pipelines seamlessly across various database platforms, ensuring data is available and consistent across systems. The setup process typically starts with defining the database sources and configuring connections to each system, including cloud-based and on-premise solutions.

In DataOps, databases are typically organized into development, testing, and production environments. Each environment is configured with specific permissions, ensuring data flows smoothly from one system to another without compromising security or integrity. By creating distinct environments, teams can avoid disruptions in production while testing new changes in isolated settings.

Moreover, setting up multiple databases in DataOps also requires defining access controls and permissions. This ensures that the right stakeholders have the appropriate level of access to each database. Access control is critical, particularly in environments that deal with sensitive data or require strict compliance with privacy regulations.

Automation plays a key role in the setup process, with DataOps tools enabling teams to deploy changes and synchronize data across databases automatically. This reduces the manual effort required for updates and ensures that all databases are aligned with the latest schema changes. The ability to automate deployment not only enhances efficiency but also helps to reduce human errors.

As part of the configuration process, defining how data will be transferred between databases is essential. DataOps platforms provide a range of integration options, such as batch processing, real-time streaming, and API-based integrations, allowing teams to choose the best method for their use case. This flexibility ensures that data is transferred most efficiently depending on each workflow's requirements.

Finally, monitoring and managing these databases effectively is essential for maintaining their health and performance. Setting up monitoring tools that track database activity, performance metrics, and potential issues helps teams to identify problems early and take corrective action before they impact business operations. Proper monitoring, combined with alerting and logging, ensures that database performance remains optimal as the number of databases increases in complex environments.

Optimizing Cross-Database Workflows

Optimizing cross-database workflows in DataOps requires careful planning and efficient execution across multiple systems. One of the primary considerations is data synchronization. When dealing with various databases, ensuring that data is consistently and accurately transferred between them is crucial for maintaining a seamless workflow. A well-designed synchronization strategy helps prevent data discrepancies and ensures that all systems reflect the same data at any time. This is especially important when the databases are part of a complex data pipeline that spans multiple teams and tools.

Teams should focus on implementing robust integration strategies to optimize cross-database workflows. Real-time streaming, batch processing, and event-driven triggers ensure data can be moved between systems quickly. Real-time integrations, in particular, are critical when working with applications that require up-to-date data. Automation tools within DataOps platforms help streamline this process by eliminating manual intervention and reducing the risk of errors during data transfers.

Access control and security policies also play a key role in optimizing cross-database workflows. Each database in the workflow may have different security requirements, and managing access across multiple systems can be challenging. Organizations can ensure that only authorized personnel can access sensitive data and configurations by implementing centralized access controls and monitoring solutions. Additionally, encrypting data in transit and at rest across databases helps protect against security breaches and ensures compliance with regulatory standards.

Another factor in optimizing workflows is ensuring that databases are correctly indexed. Efficient indexing strategies help speed up query execution across different databases. When data is queried across multiple systems, poorly designed indexes can create bottlenecks, slowing down the entire pipeline. Proper indexing also reduces the load on databases and prevents excessive resource consumption. By continuously monitoring and optimizing index performance, teams can improve query speeds and minimize operational costs.

Error-handling and recovery processes are equally crucial for optimizing cross-database workflows. Since data pipelines often involve data transfer across multiple environments, the likelihood of encountering errors is higher. Designing a workflow that can automatically detect and recover from mistakes ensures that workflows run smoothly without manual intervention. This can involve implementing retries, fallback mechanisms, and detailed error logs to identify and resolve any issues quickly.

In addition to these technical aspects, optimizing cross-database workflows also requires fostering collaboration between teams. Different teams may be responsible for other databases or parts of the pipeline, and maintaining clear communication is essential for successful collaboration. By adopting a shared data governance framework, teams can align their efforts and ensure that data is being transferred, transformed, and consumed in a standardized and consistent manner. This level of collaboration increases efficiency and reduces the chances of conflicting workflows or issues arising during data integration.

Finally, continuous monitoring and performance tuning are vital components of an optimized cross-database workflow. Regular monitoring ensures that performance issues can be detected early, allowing teams to address them proactively. DataOps platforms typically provide built-in analytics and monitoring tools that track data flow, system health, and performance metrics across databases. These tools help teams identify potential bottlenecks or inefficiencies in the workflow and take corrective action, resulting in a more efficient and reliable system.

Best Practices and Troubleshooting

Best practices for managing multiple databases in a DataOps environment are critical for ensuring efficiency, security, and scalability. One primary principle is standardizing data practices across all databases. This includes consistent naming conventions, standard data formats, and unified schema designs to reduce complexity and simplify integration. When data is standardized, data quality is maintained throughout the process, making it easier for teams to access, manipulate, and analyze data across various platforms.

Another key best practice is ensuring all databases are monitored in real time. Continuous monitoring allows teams to identify and address issues before they escalate. This includes monitoring the performance of queries, data load times, and potential bottlenecks. Leveraging automated alerts and performance metrics helps track the health of each database in the pipeline and identify any areas requiring optimization. For instance, slow queries or data transfer errors can be flagged early, allowing teams to take corrective actions before they impact the workflow.

Data security should also be considered a top priority. When managing multiple databases, it's essential to implement centralized access controls and ensure that only authorized users can interact with sensitive data. Encrypting data in transit and at rest across databases adds an extra layer of protection. Compliance with regulatory standards such as GDPR or HIPAA must also be maintained to prevent any legal or reputational risks. By enforcing tight security protocols, organizations can reduce the chance of data breaches while fostering trust with stakeholders.

The integration of version control systems is another best practice in multi-database environments. Version control allows teams to track changes in configurations, schema, and data pipelines, ensuring that updates are made in a controlled and reversible manner. This reduces the risk of introducing errors into the system that could propagate across databases. It also makes it easier to roll back to previous configurations in case of issues, preserving the integrity of the workflow. By integrating version control, teams can better manage the evolution of their database environments over time.

Effective troubleshooting is a critical aspect of maintaining a smooth-running multi-database system. In many cases, issues will arise due to inconsistent data or integration failures, and knowing how to diagnose and resolve these problems efficiently can save significant time and resources. One common troubleshooting step involves reviewing logs for errors related to data transfers or query failures. Ensuring that these logs are detailed and accessible allows teams to quickly pinpoint the root causes of issues. For example, if data is not syncing between databases, the logs might show failed connections, incorrect query formats, or conflicts in data schema that require attention.

When troubleshooting cross-database workflows, breaking down the workflow into smaller components and isolating potential problem areas is essential. Start by verifying that each database connection is functioning correctly. Once connections are confirmed, focus on the specific data transfer mechanisms between databases, looking for discrepancies in how data is mapped, formatted, or loaded. By isolating each step, you can quickly identify the specific point at which the workflow fails. This systematic approach ensures that problems are addressed logically and efficiently.

Another consideration in troubleshooting is maintaining good communication between the different teams managing the databases. Issues often arise when changes made in one database are not communicated to the other teams or there is a lack of coordination in data governance practices. Keeping all teams in the loop and ensuring that documentation is up-to-date can prevent many errors from occurring in the first place. Collaborative troubleshooting, involving cross-functional teams, often leads to quicker resolutions and helps identify recurring issues that may require long-term solutions.

Finally, developing a robust testing framework for database integrations can help prevent issues before they arise. By setting up automated tests for each step in the workflow, from data ingestion to transformation and loading, teams can verify that everything is working as expected. Running tests in isolated environments or sandboxes before deploying changes to production ensures that new configurations do not disrupt existing systems. Regularly testing configurations and integrations will help improve system reliability, reduce downtime, and simplify troubleshooting when problems occur.

Conclusion

Managing multiple databases effectively in a DataOps environment is essential for ensuring data quality, security, and operational efficiency. Teams can streamline their database management process by following best practices such as standardized configurations, real-time monitoring, and enforcing robust security measures. Integrating version control and automated testing frameworks further enhances control and reduces potential disruptions across systems.

Troubleshooting cross-database workflows requires a methodical approach, identifying and isolating problem areas in the integration process. Clear team communication and updated documentation are critical in preventing recurring issues. Leveraging detailed logs and systematic isolation can significantly reduce downtime, enabling quick resolutions.

Additionally, ensuring collaboration and proper testing strategies before changing production environments helps maintain the system's overall health. Effective management and proactive measures allow organizations to scale their data environments seamlessly without sacrificing performance or security.

With these best practices, teams can achieve greater stability, efficiency, and adaptability across multiple databases. Regular audits and workflow improvements also allow teams to anticipate potential issues, enhancing the long-term sustainability of data operations.

CHAPTER 20

DataOps.live Orchestrators

DataOps.live Orchestrators provide a central platform for automating and managing data flow within a DataOps pipeline. Orchestrators help streamline workflows by enabling data processing, transformation, and scheduling tasks. With a focus on scalability, these orchestrators allow teams to coordinate multiple pipelines, ensuring seamless integration between systems and enhancing collaboration.

Integrating orchestration tools simplifies complex workflows, making it easier to maintain and adjust processes in real time. It also improves operational efficiency by minimizing manual interventions and reducing the risk of human error. With the right orchestration strategy, teams can accelerate data pipelines, quickly address bottlenecks, and optimize resource usage.

Moreover, DataOps.live Orchestrators facilitate monitoring and troubleshooting by providing detailed insights into pipeline execution. Teams can track each step, gaining visibility into potential issues before they affect production environments. This real-time feedback loop ensures quick detection and resolution of errors, contributing to overall system reliability.

As part of the broader DataOps ecosystem, orchestration tools are pivotal in bridging the gap between development, operations, and data teams. They allow for better coordination and ensure that different elements of a pipeline are executed in the proper sequence. By optimizing this flow, organizations can ensure more accurate data processing and faster insights.

Incorporating orchestration into DataOps practices also allows for the automation of routine tasks, improving consistency across data workflows. This reduces the time spent on repetitive tasks and frees up resources to focus on higher-value activities, such as data analysis and innovation. Orchestrators thus empower teams to focus on the core objectives of their data strategies.

R. L. Steelman, Jr., *Mastering Snowflake DataOps with DataOps.live*, https://doi.org/10.1007/979-8-8688-1754-0_20

This chapter will explore the core concepts, functionality, and best practices for using DataOps.live Orchestrators to enhance data workflow management. Understanding how to integrate orchestration effectively into your DataOps processes can lead to more efficient, resilient, and scalable data operations.

Data.World Orchestrator

The Data.World Orchestrator is a key component of the DataOps.live ecosystem, designed to streamline the process of managing and automating data workflows. It acts as a central point for orchestrating data movement across different platforms, ensuring that data is processed efficiently and accurately at every pipeline stage. By leveraging this tool, teams can avoid bottlenecks and ensure that their data operations run smoothly without needing constant manual intervention.

One of the most significant advantages of the Data.World Orchestrator is its ability to integrate with various data sources and tools seamlessly. This integration capability allows it to work with cloud-based and on-premises systems, making it versatile and adaptable to multiple business environments. Whether you're working with structured, semi-structured, or unstructured data, the orchestrator can handle it, providing a unified platform for data workflows.

The orchestrator also plays a vital role in automating complex workflows. For example, it allows data teams to schedule tasks such as data extraction, transformation, and loading (ETL) at specific intervals. This level of automation helps reduce the risk of human error, ensuring that data is consistently processed and available when needed. Additionally, automation improves operational efficiency by minimizing manual oversight and streamlining the data flow process.

Another key feature of the Data.World Orchestrator is its monitoring and alerting functionality. It provides real-time visibility into the status of your workflows, allowing teams to track the progress of data pipelines and quickly identify any potential issues. If a problem arises, such as a failed task or a delay in data processing, the orchestrator can trigger automatic alerts, enabling teams to take corrective action before any significant disruption occurs.

The flexibility of the Data.World Orchestrator extends to its ability to handle complex data pipelines with multiple dependencies. Many data workflows involve interconnected tasks that must be completed in a specific order. The orchestrator's advanced dependency management ensures that each task is executed in the correct sequence, preventing any errors arising from mismanaged data flows. This feature is handy for organizations with complex data ecosystems that involve multiple teams and tools.

Moreover, the Data.World Orchestrator supports version control, which is essential for managing the evolution of data workflows over time. As data operations grow and evolve, it's crucial to be able to track changes and roll back to previous versions if necessary. The orchestrator's version control feature ensures that teams can maintain a clear history of their workflows, making auditing changes and troubleshooting issues more manageable.

One notable feature of the Data.World Orchestrator is its scalability. Whether your organization processes a few gigabytes of data or petabytes, the orchestrator can scale to meet your needs. This scalability ensures that the orchestrator will continue to function efficiently as your data operations grow, allowing your team to focus on other aspects of data management rather than worrying about performance limitations.

Collaboration is another key benefit that the Data.World Orchestrator enables. Coordination and communication can become challenging with multiple teams working on different parts of the data pipeline. The orchestrator provides a collaborative environment where teams can share workflows, monitor tasks, and resolve issues. This improves overall efficiency and helps avoid silos within the organization.

In terms of security, the Data.World Orchestrator incorporates robust measures to protect sensitive data. It supports role-based access control, allowing administrators to set permissions and define who can access specific workflows or data. This granular level of control ensures that only authorized personnel can make changes to the pipeline or view sensitive information, thereby reducing the risk of data breaches.

The orchestration platform also includes features to ensure data quality. Maintaining data integrity is essential as data flows through various pipeline stages. The Data.World Orchestrator provides built-in validation and cleansing tools that can be used to check the quality of the data at each step of the process. This helps maintain accurate and reliable data, essential for making informed business decisions.

Another important consideration is the support for both batch and real-time processing. Many organizations require real-time data processing to enable time-sensitive decision-making, while others rely on batch processing for large-scale data transformation. The Data.World Orchestrator supports both processing modes, allowing teams to choose the best approach for their specific use case.

The user interface of the Data.World Orchestrator is designed to be intuitive, making it easier for both technical and non-technical users to manage workflows. Whether you're a data engineer, a business analyst, or a data scientist, the orchestrator provides a simple and effective interface for interacting with the system. This user-friendly design reduces the learning curve and allows teams to start quickly.

Lastly, the Data.World Orchestrator is built with compliance in mind. It adheres to industry standards and best practices for data governance, ensuring that organizations can meet regulatory requirements. This is particularly important for industries such as healthcare and finance, where data privacy and security are critical.

In summary, the Data.World Orchestrator is a powerful tool for managing data workflows, offering a wide range of features designed to improve efficiency, scalability, and collaboration. Its integration capabilities, monitoring tools, and support for version control make it a valuable asset for teams working with complex data environments. Whether you want to automate processes, monitor data flows, or ensure compliance, the Data.World Orchestrator can help optimize your data operations.

VaultSpeed Orchestration in DataOps.live

The VaultSpeed Orchestrator is a vital tool in the DataOps ecosystem, designed to simplify and automate data modeling, transformation, and integration. VaultSpeed helps data teams by creating efficient data pipelines and workflows, significantly reducing the complexities of building and maintaining data models. This orchestration tool is particularly beneficial for teams looking to streamline their data management processes and improve the scalability of their operations.

One of VaultSpeed's key features is its ability to integrate seamlessly with existing data platforms, allowing organizations to automate many data workflows. It supports both cloud-based and on-premises environments, making it versatile and adaptable to various business needs. This means that whether a company uses modern cloud services or traditional data systems, VaultSpeed can help unify and automate data management.

Another essential aspect of VaultSpeed is its focus on reducing manual efforts in the data modeling process. Data engineers can use the tool to automate the creation of data models, which are typically time-consuming and error-prone when done manually. VaultSpeed leverages metadata-driven automation to create accurate, repeatable models that ensure consistency and reliability throughout the data pipeline.

VaultSpeed also excels at optimizing data pipelines for speed and efficiency. The orchestrator helps ensure data transformations are executed optimally, reducing unnecessary computations and improving overall performance. This optimization is particularly critical in large-scale data environments, where the volume of data can slow down processing times if not managed efficiently.

For teams working with complex data models, VaultSpeed provides advanced features for managing dependencies between various data elements. This helps ensure that changes made in one part of the data pipeline do not inadvertently cause issues elsewhere. The orchestrator's dependency management capabilities are essential for ensuring data integrity and avoiding problems such as data duplication or inconsistencies.

Additionally, VaultSpeed integrates with various other tools in the DataOps ecosystem, allowing for a more cohesive and interconnected approach to data operations. Whether interacting with other orchestration tools, data quality solutions, or monitoring systems, VaultSpeed ensures that all parts of the data pipeline work together seamlessly. This holistic approach helps eliminate silos and provides teams with a comprehensive view of their data workflows.

VaultSpeed also simplifies the process of handling complex data governance tasks. With data privacy and compliance becoming more critical in various industries, the orchestrator helps ensure data management processes comply with relevant regulations. By automating governance tasks, VaultSpeed reduces the risk of human error and ensures that data is handled following organizational and regulatory standards.

In terms of scalability, VaultSpeed is designed to grow with an organization's needs. As data volumes increase and workflows become more complex, the orchestrator adapts to ensure high performance. This scalability ensures that VaultSpeed remains an effective tool as organizations scale their data operations, whether expanding their data infrastructure or incorporating new data sources.

Security is also a top priority for VaultSpeed. The tool provides robust security measures to protect sensitive data throughout the entire pipeline. Role-based access control (RBAC) ensures that only authorized personnel can access or modify certain parts of the data pipeline. This level of security is crucial for organizations handling sensitive or confidential data.

Regarding user experience, VaultSpeed offers an intuitive interface that makes it easy for technical and non-technical users to interact with the tool. The user interface is designed to be accessible while providing advanced functionality for data engineers and architects. This ensures that the tool can be used effectively across teams with varying levels of technical expertise.

Another critical aspect of VaultSpeed is its ability to automate data lineage tracking. With data pipelines becoming increasingly complex, understanding data flow from source to destination is essential. VaultSpeed automatically tracks data lineage, providing visibility into how data moves through the pipeline. This visibility is essential for debugging, auditing, and ensuring data accuracy.

VaultSpeed also supports integration with a variety of data sources, which helps organizations create comprehensive data pipelines that draw from multiple platforms. Whether you are working with relational databases, NoSQL systems, or cloud-based data lakes, VaultSpeed can handle diverse data sources, ensuring that all your data is integrated into a unified pipeline.

For teams looking to enhance their data quality, VaultSpeed provides tools to help automate data validation and cleansing. As data moves through the pipeline, ensuring it meets certain quality standards is crucial. VaultSpeed allows teams to set up rules and validations to ensure that only clean, accurate data makes it to the end of the pipeline. This feature helps improve the reliability of the data and the insights derived from it.

Lastly, VaultSpeed provides a robust set of reporting and monitoring tools. Teams can monitor the status of their data workflows in real time and receive alerts when issues arise. This proactive monitoring ensures that potential issues are identified and addressed before they escalate, helping maintain the integrity and performance of the data pipeline.

VaultSpeed is an essential tool for any organization looking to streamline its data orchestration process. Its powerful features, including automation, scalability, and security, make it valuable to any data operations environment. Whether you are building complex data models or managing large-scale data workflows, VaultSpeed provides the tools and capabilities needed to optimize and automate your data operations, ensuring that your data is accurate, secure, and easily accessible.

Conclusion

While VaultSpeed provides a comprehensive orchestration solution, it's important to recognize that numerous other orchestrators are available, each with its own strengths and tailored use cases. Platforms such as Apache Airflow, Prefect, and Dagster offer alternatives for building data workflows, each emphasizing scalability, flexibility, and integration with various data tools. Choosing the right orchestrator depends on your organization's needs, team expertise, and infrastructure.

Building a custom orchestrator is also an option for teams seeking complete control over their orchestration environment. This approach allows for creating workflows tailored precisely to your organization's requirements, from data modeling to transformations and integrations. While building your own orchestrator requires significant development resources and expertise, it offers the ultimate flexibility in adapting to unique workflows and business logic.

That said, deciding between using a pre-built orchestrator or developing your own depends on several factors. Pre-built solutions like VaultSpeed provide ready-made features that can save time and reduce complexity, making them ideal for teams that need to deploy a reliable, tested solution quickly. On the other hand, a custom-built orchestrator can evolve as the organization's needs change, offering adaptability for complex or rapidly changing environments.

One key advantage of working with existing orchestrators is the support community. Established tools like VaultSpeed, Apache Airflow, and Prefect have extensive documentation and user communities that can assist with troubleshooting and feature enhancements. Leveraging these communities can significantly shorten development cycles and provide valuable insights into best practices.

Combining multiple orchestrators may also benefit larger enterprises or organizations dealing with complex data pipelines. Orchestration tools for different segments of the data lifecycle, such as ETL pipelines, data quality, and model training, can enable a more modular, specialized approach. In such cases, integrating these orchestrators through a unified control plane ensures smooth cross-platform operations.

Ultimately, whether you opt for a solution like VaultSpeed or build your own orchestrator, the important takeaway is the value of orchestration in ensuring data workflows are optimized, reliable, and scalable. The right orchestration strategy ensures that teams can focus on deriving insights and value from data, rather than managing the complexities of data movement, transformation, and integration. By carefully considering the options, organizations can choose or create the orchestration solution that best aligns with their goals.

CHAPTER 21

Build Only Changed Models

In data operations, optimizing build processes is critical to maintaining efficiency, especially when dealing with large-scale models. Building only the changed models ensures that workflows are faster, resources are conserved, and the overall process becomes more efficient. By building models incrementally, data teams can focus on areas that have been modified, avoiding the unnecessary rebuilding of unchanged models.

This chapter will focus on how to set up and manage the incremental build process for MATE models, a practice that can significantly improve performance. We'll explore how this strategy fits into the more extensive data pipeline and the role of MATE in streamlining the data transformation process. With this approach, you can target specific areas of the model that need attention, reducing downtime and computational costs.

Additionally, we'll discuss best practices for implementing this method, address common pitfalls, and discuss how to manage dependencies effectively. By the end of this chapter, you'll be equipped with the knowledge to configure your pipeline to build only changed models, improving your team's productivity and optimizing resource utilization in data operations.

Building Change-Only Models

To configure the "Build Only Changed Models" feature in MATE, you begin by enabling the change detection capability within your project configuration. This feature identifies which models have been modified since the last build, allowing MATE to focus on those specific models rather than rebuilding the entire set each time. The change detection mechanism is key to improving build times and optimizing computational resources.

R. L. Steelman, Jr., *Mastering Snowflake DataOps with DataOps.live*, https://doi.org/10.1007/979-8-8688-1754-0_21

The configuration of change detection typically involves setting up the change_detection flag in your pipeline configuration. This flag ensures that only the modified models are built by comparing the current state of the models to the previous state stored in the metadata. Doing so helps minimize unnecessary processing and avoids redundant builds.

Next, you need to define the granularity of change detection. Depending on your project, you can configure change detection on the individual model level or based on broader dependencies. The level of granularity can be adjusted to ensure that even models with slight alterations trigger a rebuild, while avoiding over-building when changes are minimal.

Leveraging version control systems (VCS) is essential for a more accurate change detection setup. MATE integrates seamlessly with Git, enabling you to track changes granularly. This integration allows MATE to monitor and detect any changes in the source code and data schema that could affect the output of your models. This ensures that all modifications are captured without missing any critical updates.

Additionally, MATE provides tools to handle dependencies effectively. Since models in data pipelines often have interdependencies, it's crucial to ensure that only the changed models and their dependent models are rebuilt. This prevents unnecessary rebuilds of models that have not been modified but rely on other models that have. The system manages these dependencies automatically, optimizing the rebuild process.

Consider incorporating conditional logic into your build process to improve build efficiency further. This logic can control when certain models should be built based on specific triggers, such as data changes or updates to external dependencies. By incorporating conditions like these, you can avoid rebuilding models irrelevant to the latest data changes, enhancing both speed and resource utilization.

When configuring change detection, it is essential to account for different environments. Your development, staging, and production environments may require different configurations for change detection. MATE allows you to set up distinct pipelines for each environment, ensuring that models are built only when necessary based on changes specific to that environment.

Finally, while configuring "Build Only Changed Models" can save time and resources, monitoring the system's performance is essential. Regular audits and testing of the change detection setup can help identify potential issues or misconfigurations. Ensuring that the change detection logic is working as expected will ensure your pipeline remains efficient and reliable, minimizing downtime and unnecessary builds.

CHAPTER 22

DataOps.live REST API

The DataOps.live REST API offers users a powerful and flexible way to interact programmatically with the DataOps.live platform. The API enables seamless automation, integration, and custom workflows across various applications by exposing key platform functionalities through RESTful endpoints. The ability to interact with multiple components such as projects, environments, and pipelines ensures that users can control their DataOps workflows and orchestrations directly from external systems.

Authentication is a crucial component of using the DataOps.live REST API, which employs token-based authentication to ensure that only authorized users can access the API. When interacting with the API, users must generate a personal access token (PAT), which is then passed in the header of HTTP requests. This token serves as proof of identity and access permissions, ensuring secure and reliable interactions with the platform.

DataOps.live's REST API is organized around the platform's resource model. Each resource, such as a pipeline or project, has its own endpoints supporting different operations like creating, updating, and deleting resources. For instance, users would send a POST request to the respective endpoint to create a new project, passing the required parameters in the request body. These resource-centric endpoints allow for precise and intuitive management of each component in the DataOps lifecycle.

One of the most powerful aspects of the DataOps.live REST API is the ability to interact with pipeline operations. Pipelines in DataOps.live can be managed, monitored, and triggered programmatically. Users can start, stop, or monitor the execution of pipelines through specific endpoints, giving them complete control over their automated workflows. This enables advanced automation scenarios where external systems or scheduling platforms can trigger DataOps pipelines as part of more extensive data processes.

The DataOps.live REST API also supports integration with third-party services and tools. Whether you want to integrate with a CI/CD platform like Jenkins or an external data warehouse, the REST API makes it possible to connect and exchange data between

R. L. Steelman, Jr., *Mastering Snowflake DataOps with DataOps.live*, https://doi.org/10.1007/979-8-8688-1754-0_22

platforms. For instance, you can configure the API to pull in data from external systems or send data back to them, creating a seamless flow of information between DataOps.live and other parts of your infrastructure.

API responses are typically returned in JSON format, making it easy for developers to parse and handle the data. This format is widely used and understood across various programming languages, ensuring users can easily integrate the API into their existing systems without significant overhead. The responses include data about the resources created, modified, or retrieved, along with status information to indicate success or failure of operations.

Error handling is another key aspect of the API. If a request fails, the API returns an appropriate error code and message, providing users with the necessary information to troubleshoot and resolve issues. Common errors include authentication failures, invalid input, or unavailable resources. By providing detailed error messages, the API ensures that users can quickly diagnose issues and take corrective actions to restore functionality.

The API provides support for batch processing for users seeking to automate or integrate DataOps workflows. For example, users can send multiple requests in a single API call or programmatically execute a sequence of tasks like triggering a series of pipelines, monitoring their status, and collecting the results. This ability to handle batch operations improves efficiency and reduces the amount of manual work required for large-scale processes.

The DataOps.live REST API documentation offers detailed information on endpoints, request parameters, and response formats. It includes example requests and responses, making it easier for developers to understand how to interact with the platform. This documentation is essential for developers and administrators, providing a comprehensive guide to utilizing the API's full potential.

Another significant feature of the DataOps.live REST API is its ability to manage environment configurations. Users can create, modify, and delete environments by sending requests to the relevant endpoints. This functionality is crucial in multi-environment setups where data needs to be validated across different stages, such as development, testing, and production. The API simplifies the management of these environments, reducing the effort required to maintain multiple configurations.

The API also includes endpoints for managing users and permissions. Administrators can use the API to manage access control, ensuring only authorized users have the necessary permissions to interact with specific projects, pipelines, or environments. This functionality is essential for organizations that require strict governance and security over their DataOps processes.

In addition to managing individual components, the API provides functionality for automating workflows. With the ability to start, stop, and schedule pipeline executions, users can fully automate their data processes. This is particularly useful in environments where data must be processed at regular intervals or in response to specific triggers, such as new data arriving in a source system.

To enhance the flexibility of the DataOps.live REST API, the platform supports webhooks for real-time notifications. Webhooks enable users to receive event-based updates, such as when a pipeline starts, stops, or encounters an error. This allows external systems to respond to changes in real-time, facilitating integrations with monitoring tools or alerting systems.

The REST API also supports versioning, ensuring that updates to the platform do not break existing integrations. The versioning system allows users to specify which version of the API they wish to interact with, providing stability for long-term projects and integrations. Users can take advantage of new features as the platform evolves while maintaining compatibility with older versions.

Finally, the DataOps.live REST API is designed for scalability. Whether you're managing a small project or a large-scale data pipeline, the API can handle high volumes of requests without compromising performance. This makes it suitable for both small teams and enterprise-level organizations, providing a reliable foundation for automating and managing DataOps workflows at scale.

CHAPTER 23

Medallion Architecture

Medallion Architecture is a modern data architecture framework designed to optimize data flow through different layers of transformation. It's a layered approach that provides a clear structure for managing data pipelines, ensuring that raw data is efficiently processed and enriched into usable formats for analysis. The architecture typically involves three core layers: Bronze, Silver, and Gold, each serving a distinct role in transforming raw data into valuable insights.

This chapter will explore how Medallion Architecture improves data operations by introducing structure and governance. Organizations can ensure data quality, integrity, and scalability by breaking down the data pipeline into clear stages. We will also examine how Medallion architecture allows teams to manage and monitor the entire lifecycle of their data more effectively, promoting better collaboration between different stakeholders.

By the end of this chapter, you will have a comprehensive understanding of how Medallion Architecture fits into modern data ecosystems. You will learn the specific role of each layer in transforming data and how to implement it within your own data environment, especially when working with advanced tools like DataOps.live and Snowflake.

Understanding Medallion Architecture: Key Concepts and Principles

Although Medallion Architecture has been around for a long time, it remains a tried and proven framework that remains highly relevant in modern data environments. Its approach of segmenting data processing into distinct layers has stood the test of time due to its simplicity, scalability, and efficiency (Figure 23-1). The architecture has been widely adopted across industries because it addresses many challenges faced by data

R. L. Steelman, Jr., *Mastering Snowflake DataOps with DataOps.live*, https://doi.org/10.1007/979-8-8688-1754-0_23

teams today, such as data quality, governance, and the need for adaptable workflows. Despite the emergence of new technologies and frameworks, Medallion's robust structure ensures it remains a viable and effective solution for managing data pipelines.

Figure 23-1. *Medallion Architecture*

Medallion Architecture is a modern approach to data modeling that organizes data in multiple layers to make it more manageable, accessible, and meaningful for business analysis. The core concept revolves around breaking down the data pipeline into distinct stages, each with its role in processing and transforming data. The architecture is commonly divided into three layers: Bronze, Silver, and Gold. Each layer handles a different aspect of data processing, and together, they form a well-organized flow that supports a wide range of analytics and business intelligence use cases.

In the Bronze layer, raw data is ingested from various sources and stored in its most unrefined form. This stage focuses on capturing data in a way that preserves its integrity, ensuring that it remains unchanged for auditing and future analysis. The Bronze layer typically contains a mix of structured, semi-structured, and unstructured data. It's designed to handle all types of incoming data without needing transformation or cleaning, providing a solid foundation for future steps.

The Silver layer takes over once the raw data is in the Bronze layer. This layer is where data transformation and cleansing occur. Data engineers or analysts use this stage to perform necessary operations such as data cleaning, formatting, and enrichment. This stage can also include the application of business rules, allowing users to transform data into more usable formats. The Silver layer is crucial for ensuring the data is reliable and ready for more advanced analysis in the next stage.

The Gold layer is the final stage in the Medallion Architecture and represents the highest data quality. In this layer, the data is aggregated, summarized, and structured into formats that are directly useful for business decision-making and analytics. Data in the Gold layer is typically optimized for performance, focusing on creating business intelligence-ready datasets that can be used for reporting, dashboards, and advanced analytics. This layer generates insights, making it the most valuable stage in the architecture.

A critical principle of Medallion Architecture is its iterative nature. Data flows through the architecture in stages, with each layer adding value to the data and making it more refined and actionable. This process allows for flexibility and adaptability, as new data sources can be incorporated into the system and processed consistently and repeatedly. The iterative flow ensures that data quality improves progressively, and insights become more precise as it moves through each layer.

Medallion Architecture also emphasizes modularity. The distinct separation of the Bronze, Silver, and Gold layers allows organizations to manage and maintain their data pipelines more effectively. Each layer can be scaled independently, which helps manage large data volumes and ensures that processing tasks can be distributed across different systems or teams. This modularity provides greater control over the data lifecycle and reduces the risk of errors by isolating data processing tasks.

Another key aspect of Medallion Architecture is its adaptability to various data environments. Medallion Architecture provides a flexible framework for organizing data processing pipelines, whether organizations work with on-premises systems, cloud platforms, or hybrid environments. This adaptability also extends to different data types, including structured, semi-structured, and unstructured data, making it suitable for a wide range of use cases, from operational reporting to machine learning.

The Medallion model supports agility, which is essential for modern data operations. As data environments evolve and business requirements change, Medallion Architecture allows organizations to quickly adapt by adding new data sources, adjusting transformations, or refining business rules. This agility ensures that data teams can respond to new demands with minimal disruption, and the architecture can scale as the organization grows.

Another advantage of the Medallion Architecture is its alignment with modern data processing frameworks and tools. Many organizations use tools such as Snowflake, Databricks, or Apache Spark, which are well-suited for implementing Medallion Architecture. These tools support large-scale data processing, and their integration with the architecture allows teams to automate tasks, streamline workflows, and reduce manual effort in the data pipeline. This integration further enhances the architecture's efficiency and scalability.

Finally, Medallion Architecture supports better governance and compliance. The clear separation of layers and the ability to trace data as it moves through each stage make implementing policies around data security, privacy, and auditing easier. By keeping raw data in the Bronze layer and applying transformations in a controlled environment, organizations can maintain high data governance standards while complying with regulatory requirements.

The Bronze Layer: Raw Data Ingestion and Storage

The Bronze layer is the first stage of the Medallion Architecture, where raw data is ingested and stored in its most unrefined form. At this point, data is gathered from various sources, such as databases, logs, sensors, APIs, and third-party systems. The key goal is to preserve the integrity of the data, maintaining it in its original structure without any transformation or cleaning. This step ensures that no data is lost and accurately represents the raw inputs, serving as a foundation for later stages.

Data can come in many formats in this phase, including structured, semi-structured, and unstructured. Structured data, such as tables and relational datasets, is easy to manage and analyze, while semi-structured data, like JSON or XML files, requires more effort for parsing and processing. Unstructured data, such as images, videos, or logs, may need further transformation before it can be helpful in analysis. The Bronze layer is designed to handle all these data types, ensuring that even the most complex datasets can be captured.

One of the key features of the Bronze layer is its focus on data preservation. By keeping the data in its original form, organizations can ensure that they have an immutable record of all incoming information. This is particularly important for regulatory compliance and auditing purposes, as it allows teams to trace the data's lineage and verify that it has not been altered or corrupted. In addition, preserving raw data means that it can be revisited and transformed later if needed, which offers flexibility in the long run.

Data in the Bronze layer is typically stored in a highly scalable storage system, such as a data lake or cloud-based storage solution. This storage system is designed to accommodate large volumes of data, ensuring that data is stored efficiently and can be accessed quickly for future processing. Cloud platforms like AWS, Azure, or Google Cloud provide the infrastructure to scale the storage as data volumes grow. These platforms also offer features such as automatic data replication and fault tolerance, ensuring data availability and security.

The raw data in the Bronze layer is typically organized in its most basic form, often partitioned by time, source, or type to make it easier to manage. However, little to no business logic or data governance policies are applied at this stage. This makes the Bronze layer highly flexible and adaptable to various data sources. It serves as the initial "catch-all" for data before any processing or transformation begins, allowing data engineers to focus on ensuring that all incoming data is properly captured without worrying about structure.

Another important consideration in the Bronze layer is data quality monitoring. While no transformations are applied at this stage, organizations often set up monitoring mechanisms to ensure the data being ingested is accurate and complete. These monitoring systems can check for missing data, outliers, or inconsistencies that may indicate issues with the source systems or data collection processes. This early-stage quality assurance helps identify problems quickly, allowing teams to take corrective action before data is moved to the next layer.

Data governance also plays a vital role in the Bronze layer, especially when dealing with sensitive information. Data governance practices such as access control, encryption, and auditing are implemented to ensure that only authorized users can access the raw data and that it is protected from unauthorized tampering. By applying these practices early in the data pipeline, organizations can ensure that they adhere to compliance standards and protect customer and business data from security risks.

Finally, one of the main advantages of the Bronze layer is its ability to support historical data analysis. Since raw data is preserved in its entirety, organizations can revisit historical datasets whenever needed, whether for compliance reporting, troubleshooting, or deep analysis. This is particularly valuable for industries that require long-term data retention and historical trend analysis. Organizations can easily perform time-series analysis, trend identification, and other historical analyses by maintaining a complete record of all data.

Transforming Data in the Silver Layer: Cleaning and Enrichment

In the Silver layer of the Medallion Architecture, the focus shifts from raw data ingestion to data transformation. This is the stage where data cleaning and enrichment occur, refining the data collected in the Bronze layer and preparing it for more complex analysis. The Silver layer is an intermediary between raw data and business-ready insights, offering structured, standardized, and enriched data to facilitate decision-making.

The primary goal of the Silver layer is to address data quality issues present in the raw datasets. This can involve filtering out duplicate records, correcting inaccuracies, or filling in missing values through methods such as imputation or data interpolation. These transformations help ensure that the data in the Silver layer is reliable and ready for more detailed analysis in later stages of the data pipeline.

Data enrichment also plays a significant role in the Silver layer. Here, data can be enhanced by integrating additional sources or applying business logic to provide more context. For instance, geographic data could be enriched by adding location-specific details, or customer data might be supplemented with external demographic or behavioral information. Enrichment adds value to the data, making it more actionable and relevant for downstream business applications.

In the Silver layer, data is often transformed into a structured format, such as tables or views, which makes it easier to query and analyze. These structures are optimized for performance and provide a transparent and standardized representation of the data, which is crucial for business intelligence tools, machine learning models, or reporting systems. By transforming the data into a usable format, the Silver layer ensures that analysts and data scientists can quickly extract insights and perform complex operations on the data.

At this stage, data cleansing techniques such as filtering, standardization, and normalization are applied to ensure consistency across different datasets. For example, textual data may be standardized to match a standard naming convention, or numerical values may be normalized to a consistent scale. These transformations reduce the risk of discrepancies and make comparing and aggregating data from various sources easier.

The Silver layer also contains more complex data processing, including business rules and logic application. This could involve calculations, such as aggregating sales figures by region or calculating customer lifetime value. These operations enrich the data and align it more with the business's needs, ensuring that the data is clean and meaningful for decision-making.

Data validation is another key process in the Silver layer. This includes verifying that the data meets predefined quality standards and conforms to specific rules or constraints. For example, ensuring that numerical fields fall within expected ranges, that date fields are valid, or that categorical values are consistent with predefined categories. These validation checks help identify and correct any issues that could lead to inaccurate or misleading insights.

The transformation process in the Silver layer can also involve creating relationships between different datasets. This is particularly important in data warehousing, where multiple tables or datasets need to be joined or linked to provide a complete view of the data. The Silver layer prepares these relationships, ensuring that data from disparate sources can be merged and analyzed meaningfully.

Depending on the organization's needs, the Silver layer often includes batch and real-time processing. While batch processing may be sufficient for periodic reporting and analytics, real-time processing can be crucial for time-sensitive applications such as fraud detection or customer personalization. By incorporating both types of processing, the Silver layer supports a wide range of use cases, from historical analysis to real-time insights.

One important consideration in the Silver layer is scalability. As data volumes continue to grow, it is essential to ensure that the processes used for data transformation can scale efficiently. This may involve optimizing algorithms, using distributed computing frameworks, or leveraging cloud-based infrastructure to ensure that transformations can be performed quickly and reliably at scale.

Finally, it is in the Silver layer that data governance practices begin to take shape. With data being transformed and enriched, it is crucial to ensure proper access controls, data lineage tracking, and auditing mechanisms are in place. These practices help ensure that the data remains secure, compliant with regulatory requirements, and traceable, providing transparency into how the data has been manipulated and used.

In summary, the Silver layer is a critical step in the Medallion Architecture, focusing on cleaning, enriching, and transforming data into a standardized format ready for analysis. This layer ensures that raw data is refined into high-quality, business-relevant datasets, paving the way for the following insights and decisions. The transformations that occur in the Silver layer are key to unlocking the value of the data, ensuring that it is accurate and actionable.

The Gold Layer: Data Aggregation and Analysis for Business Insights

The Gold layer in the Medallion Architecture is the final step in the data processing pipeline, where the data is aggregated and refined to deliver actionable business insights. This layer is the culmination of the work done in the Bronze and Silver layers, and it is designed to provide high-quality, well-structured data that can be easily consumed by business users, analysts, and decision-makers. The Gold layer focuses on delivering the most valuable and insightful data, optimized for reporting, analytics, and strategic decision-making.

In the Gold layer, data aggregation is key. This step involves combining data from different sources, transforming it into meaningful metrics, and creating summaries that provide insight into key business performance indicators (KPIs). These metrics could range from financial summaries, such as total revenue or profit margins, to operational metrics, such as customer churn rates or inventory levels. The Gold layer's data is typically organized into reports, dashboards, or data cubes that present the results in a way easily interpretable and actionable by business users.

Aggregation in the Gold layer is often done using advanced techniques like grouping, filtering, and window functions. For example, sales data may be aggregated by region, product category, or time period to identify trends and performance across different segments. Creating these summaries is crucial for businesses that need to monitor performance over time or across various business units. The Gold layer ensures that this data is aggregated consistently, accurately, and efficiently, allowing businesses to make data-driven decisions based on real-time or historical trends.

One of the key features of the Gold layer is its ability to optimize data for high-performance querying. In this layer, the data is structured to support fast queries, which is particularly important for large-scale analytics or reporting systems. Businesses rely on the Gold layer for ad-hoc analysis, drill-downs, and real-time data exploration, and the ability to quickly retrieve the most relevant data is critical for supporting decision-making in dynamic environments. To achieve this, the Gold layer often employs techniques such as indexing, partitioning, and materialized views to enhance query performance.

Data in the Gold layer is often enriched with business-specific metrics, KPIs, and performance indicators that provide meaningful insights. For example, a company's customer data in the Gold layer might include metrics like customer lifetime value, net promoter score, or average order value. These metrics add context to the data, helping businesses understand what happened and why. By applying business rules and logic to the data in the Gold layer, organizations can turn raw data into valuable insights that guide strategic decision-making.

The Gold layer often leverages advanced analytics and machine learning models to provide predictive insights and forecasts. For instance, a business might use the Gold layer to perform demand forecasting, customer segmentation, or churn prediction, all requiring sophisticated statistical techniques. These models are built on top of the data that has already been aggregated and cleaned in the previous layers, ensuring that the insights derived from the models are based on high-quality, reliable data.

Business intelligence (BI) tools and dashboards typically connect to the Gold layer for reporting and visualization. Dashboards present the aggregated data in a visual format, such as charts, graphs, and tables, making it easier for business users to understand and act on the insights. The Gold layer is where the "final product" of data processing is presented, and it is tailored to the needs of various business functions, from finance and marketing to operations and sales.

Another important aspect of the Gold layer is data governance. While the Bronze and Silver layers focus on ensuring data quality, the Gold layer emphasizes ensuring that the data is aligned with the organization's strategic goals. This includes adhering to data privacy and security policies, managing user access to sensitive data, and maintaining data lineage to track the flow of data from its raw form to the final reports and dashboards. By establishing clear governance practices, organizations can ensure that the data used for decision-making is trustworthy and compliant with regulatory requirements.

The Gold layer also serves as the foundation for business collaboration. It enables teams across the organization to align their strategies and goals by providing a single source of truth for key metrics and insights. Whether the sales team uses the data to track performance against targets or the marketing team analyzes customer behavior, the Gold layer creates a common ground for data-driven collaboration. This cross-departmental alignment is essential for ensuring that all teams work toward the same objectives and use the same data to inform their decisions.

While the Gold layer is often used for reporting and analysis, it also is critical in driving automation and operational decision-making. In many modern businesses, the Gold layer feeds directly into operational systems, triggering automated actions based on predefined business rules. For instance, an e-commerce company might automate inventory reordering based on sales trends identified in the Gold layer, or a financial institution might use predictive analytics to optimize credit scoring models. By integrating the Gold layer with operational workflows, businesses can move from data-driven analysis to real-time, automated action.

To maintain the value of the Gold layer over time, organizations must continually monitor and update the data models, KPIs, and analytics within the layer. As business needs evolve, new data sources may need to be integrated, and existing models may require adjustments. Additionally, the data in the Gold layer must be refreshed regularly to ensure that it reflects the most current business conditions. By keeping the Gold layer up-to-date and aligned with business objectives, organizations can ensure that it continues to provide relevant and valuable insights.

In conclusion, the Gold layer of the Medallion Architecture is crucial for transforming cleaned and enriched data into meaningful business insights. This layer provides aggregated data, advanced analytics, and performance metrics that enable businesses to make informed decisions. By focusing on high-performance querying, predictive analytics, and data governance, the Gold layer ensures that the right data is available at the right time to support strategic decision-making across the organization.

Implementing Medallion Architecture with DataOps.live

Implementing Medallion Architecture with DataOps.live enables a structured approach to managing data workflows across multiple layers. The Bronze layer, which focuses on raw data ingestion, integrates seamlessly with the DataOps.live platform's automation tools. This makes it easier to manage the initial influx of data, ensuring that it is reliably captured and stored in the appropriate format, ready for further processing in subsequent layers.

As data flows from the Bronze layer into the Silver layer, DataOps.live facilitates data cleaning and enrichment. Through automated pipelines, users can define transformation rules that are applied to raw data, helping to maintain consistency and quality. This step in the Medallion Architecture is critical for turning unrefined raw data into something usable and trustworthy for more advanced analytics.

The Gold layer, the final stage of Medallion Architecture, benefits from DataOps.live's powerful analytics and reporting capabilities. Here, the platform enables efficient data aggregation from various sources, allowing businesses to derive actionable insights through well-organized dashboards and reports. DataOps.live's integration with BI tools allows users to access these insights in real time, helping business leaders make data-driven decisions quickly.

A key strength of DataOps.live in implementing Medallion Architecture is its ability to integrate with various data sources and applications. The platform supports integration with databases, data warehouses, and cloud storage services, which ensures that data can be consistently ingested, processed, and analyzed across the organization. This flexibility allows businesses to adapt Medallion Architecture to their existing infrastructure, facilitating smoother transitions and faster time-to-value.

One of the significant challenges in implementing Medallion Architecture is ensuring data quality at each layer. With DataOps.live, organizations can automate testing and monitoring at each stage of the data pipeline. This proactive approach to data quality ensures that any issues are detected early and corrected, minimizing the risk of poor-quality data reaching the Gold layer. Additionally, version control and auditing capabilities within DataOps.live enable teams to maintain a clear history of data transformations and track changes over time.

Collaboration is another area where DataOps.live excels in implementing Medallion Architecture. The platform supports collaboration among data teams by providing a centralized environment where users can define, manage, and share data models. This ensures that everyone from data engineers to business analysts can contribute to and access the data transformation process, aligning teams and improving the overall efficiency of data workflows.

Finally, DataOps.live allows businesses to scale their Medallion Architecture by automating the deployment and maintenance of data pipelines. With robust orchestration tools, users can automate tasks such as data refresh, model building, and transformation, freeing up time for data teams to focus on more strategic tasks. The platform's scalability ensures that the Medallion Architecture can scale seamlessly to handle increasing complexity and data requirements as data volumes grow.

Medallion Architecture Best Practices and Optimization Strategies

Medallion Architecture is a powerful framework for managing data pipelines, and implementing best practices and optimization strategies is essential for achieving its full potential. One key strategy is to ensure that the Bronze layer focuses solely on raw data ingestion, without unnecessary transformations, to maintain the integrity and authenticity of the incoming data. Avoiding over-processing at this stage ensures that you preserve the original data, which can be valuable for future use cases, especially when dealing with diverse or unstructured sources.

A best practice for optimizing Medallion Architecture is to design efficient transformation pipelines in the Silver layer. This step involves cleaning, enriching, and structuring the data, and it's critical to strike a balance between transformation complexity and performance. Avoid unnecessary transformations that might delay processing, and prioritize optimizations like partitioning and parallel processing to improve the speed and scalability of data pipelines. Incremental loads, where only new or changed data is processed, can also significantly reduce the computational load.

Another optimization technique involves the careful management of storage within the architecture. In the Bronze and Silver layers, it's important to consider the storage costs and performance trade-offs between cloud storage options, such as object storage and databases. For example, raw data in the Bronze layer can be stored in cheaper, less performance-intensive storage, while more processed data in the Silver and Gold layers should be stored in higher-performance databases for faster querying and aggregation. This ensures you only pay for the performance you need at each stage.

Automation is also key to Medallion Architecture's best practices. By automating the data pipeline as much as possible, you can eliminate manual interventions, reduce errors, and ensure more consistent results. This includes automating data transformations, quality checks, and data refreshes. In the Gold layer, this can extend to automating the generation of reports and dashboards, which ensures that the business is constantly working with the most up-to-date and relevant data without delays.

Implement robust monitoring and validation mechanisms to optimize data quality throughout the Medallion Architecture. Automated testing at each layer helps identify data quality issues early, ensuring that only high-quality, accurate data makes it to the final Gold layer. This is especially important in the Silver layer, where data transformations can introduce errors or inconsistencies. Maintaining proper data lineage and auditing capabilities also ensures transparency, making it easier to trace issues back to their source.

A common challenge in Medallion Architecture is handling large volumes of data efficiently. For this, partitioning and indexing strategies are invaluable. Partitioning divides data into manageable chunks, improving query performance and scalability, while indexing helps speed up searches within large datasets. In the Silver and Gold layers, where more complex transformations and analytics occur, leveraging advanced partitioning strategies can dramatically reduce the time it takes to process large datasets and query the results.

While optimizing for performance is important, it's equally essential to maintain flexibility in the architecture. Medallion Architecture should support incremental changes without requiring significant rework. For example, instead of rigidly defining data transformations upfront, designing the Silver layer to accommodate new data sources or transformation logic as the business evolves is beneficial. This ensures that your architecture is adaptable to new requirements, such as integrating additional data sources or changing business needs.

Collaboration and version control are also vital components of Medallion Architecture optimization. Data teams should leverage tools that support version control, such as Git, to manage changes to data models and transformation logic. This allows teams to collaborate efficiently and minimizes the risk of errors or inconsistencies when changes are made. Version control also enables teams to roll back to previous versions if a particular transformation or change causes issues in the pipeline.

In terms of specific tools and platforms, a platform like DataOps.live can be used to simplify the implementation of Medallion Architecture best practices. The platform offers built-in automation, testing, and monitoring features that allow you to build and manage data pipelines efficiently. It also integrates well with popular cloud storage providers and data warehouses, ensuring you can build a scalable and high-performance architecture without requiring extensive custom coding.

Finally, documenting your Medallion Architecture implementation is an often-overlooked but essential best practice. Proper documentation helps ensure team members understand the flow of data across the layers and the rationale behind each transformation. It also makes it easier to troubleshoot and maintain the architecture in the future. Clear documentation ensures that new team members can quickly get up to speed and contribute to the architecture, maintaining consistency as the team grows and changes over time.

By following these best practices and optimization strategies, organizations can build highly efficient and scalable Medallion Architectures, which are enabled to process, analyze, and derive insights from their data more effectively. Each of the layers—Bronze, Silver, and Gold—plays a crucial role in ensuring that data is properly ingested, cleaned, enriched, and analyzed, and optimizing each of these steps helps ensure that the architecture is as performant and scalable as possible.

Scaling Medallion Architecture for Large Data Environments

Scaling Medallion Architecture for large data environments requires a thoughtful approach to infrastructure and data management practices. As data volumes grow, organizations must ensure the architecture can handle the increased load across each layer. One key strategy for scaling Medallion Architecture is efficient partitioning. This means breaking down data into more minor, manageable pieces based on specific criteria, such as date ranges or geographical regions. Partitioning helps optimize query performance by allowing the system to process only relevant subsets of data instead of the entire dataset.

Parallel processing is another important technique for scaling Medallion Architecture. Organizations can process data more quickly and efficiently by utilizing multiple processors or nodes. This is particularly important in the Silver and Gold layers, where data is transformed and aggregated for analysis. The ability to run transformations in parallel can drastically reduce processing time, especially when dealing with large datasets. Leveraging cloud-native tools and services that support parallel processing and distributed computing can be a game-changer for scaling this architecture.

In addition to parallel processing, adopting a microservices-based architecture can further enhance Medallion's scalability. Each microservice can handle specific tasks such as data ingestion, transformation, or analysis, and these services can be scaled independently based on demand. This modular approach helps organizations scale each part of the architecture without affecting other components. It also enables more efficient resource allocation, ensuring that the system only uses the resources necessary to handle specific tasks at any given time.

Another critical aspect of scaling Medallion Architecture is optimizing data storage. In large data environments, the choice of storage solutions becomes increasingly important. Using distributed file systems or cloud-based object storage solutions allows for greater scalability, as these systems can store vast amounts of data across multiple locations and automatically scale based on demand. This ensures that the architecture remains performant and cost-effective even as data grows exponentially.

Data indexing and search optimization are also key to scaling Medallion Architecture. As the dataset grows in size, the time it takes to perform queries can increase significantly. Organizations can reduce query times by creating indexes on frequently accessed data and making the architecture more responsive. Indexing strategies should be tailored to the queries that will be executed most often, ensuring that the most relevant data is always accessible with minimal latency.

Maintaining data quality while scaling is another challenge. As the data grows in size and complexity, it becomes more difficult to ensure the accuracy and consistency of the information. Implementing automated data validation checks and quality control processes throughout the pipeline is crucial for maintaining scale-quality data. Additionally, monitoring tools that track data quality metrics can help detect and address issues early in the process, ensuring that only reliable data reaches the Gold layer for analysis.

Finally, effective monitoring and alerting mechanisms are essential for scaling Medallion Architecture. In large data environments, the risk of failures or performance bottlenecks increases. Implementing comprehensive monitoring tools that track the performance of each layer in the architecture can help detect issues before they escalate. Automated alerts can notify teams of any anomalies, allowing for rapid troubleshooting and resolution. Regular audits and performance assessments also help ensure that the architecture is continuously optimized to meet growing data demands.

In summary, scaling Medallion Architecture in large data environments requires a combination of strategies, including partitioning, parallel processing, microservices, optimized storage, indexing, data quality control, and monitoring. By carefully considering these aspects, organizations can ensure that their Medallion Architecture remains performant, scalable, and capable of handling the growing demands of big data.

Use Case: Medallion Architecture in Solar Energy

Medallion Architecture can be a robust framework for optimizing data pipelines in the solar energy industry, where vast amounts of data from various sensors, energy grids, weather forecasts, and user interactions must be processed. The Bronze layer can ingest raw data from solar panels, energy storage systems, and weather stations. Often unstructured and noisy, this data can be captured in real-time and stored in raw form without much transformation.

The Silver layer becomes essential for cleaning and enriching the data in this context. Solar energy data, for example, may have gaps, errors, or inconsistencies, such as missing time stamps, sensor malfunctions, or environmental disruptions. Transforming this raw data into a clean, structured form allows for improved accuracy in predictions and insights. Weather data, for instance, can be cross-checked with solar panel data to identify patterns and anomalies in energy production based on environmental factors like cloud cover and temperature.

In the Gold layer, the data is aggregated to provide key insights into solar energy production, usage, and performance. This could involve calculations for efficiency, performance ratios, or energy supply and demand forecasts. Aggregating and analyzing historical data can also help optimize the performance of solar energy systems over time. For example, the system might identify trends in energy production during specific months or seasons, aiding in more accurate financial planning and system maintenance.

A major use case for Medallion Architecture in solar energy is predictive maintenance. With a wealth of sensor data on individual solar panels and inverters, the architecture can be used to predict when specific units are likely to fail. Using the Silver layer, data from various sources (such as temperature or voltage readings) can be cleaned and combined with historical maintenance records to detect early signs of malfunction. Predictive models built on aggregated data in the Gold layer can forecast when a solar panel will require maintenance, thereby reducing downtime and optimizing system reliability.

Another valuable application is energy consumption forecasting. By analyzing historical data from solar energy production along with usage patterns from the Gold layer, companies can more accurately predict the energy needs of consumers or the energy grid. This information can then be used to adjust power production strategies, store energy, or optimize grid management, improving both efficiency and cost-effectiveness. In areas where solar power is a significant portion of the energy mix, this becomes a critical tool for grid operators.

Medallion Architecture also supports the integration of third-party data sources, such as weather forecasts and market demand data. In the Silver layer, weather data can be cleaned and enriched with historical solar energy production data, creating a robust system for forecasting energy output. The Gold layer then aggregates these insights to generate daily, weekly, or monthly reports to guide operational decisions, such as whether to draw power from solar panels or traditional energy sources based on real-time and forecasted conditions.

The scalability of Medallion Architecture is also key in the solar energy sector, especially as renewable energy data is often collected at scale from thousands or even millions of devices. With the Bronze layer efficiently storing raw data from a vast network of solar assets and environmental sensors, companies can scale the architecture across regions without worrying about overwhelming their data storage infrastructure. The Silver and Gold layers can be scaled horizontally to handle large amounts of data aggregation and transformation, ensuring that insights can be derived at any scale.

Additionally, Medallion's flexibility allows solar energy companies to integrate different data types. For instance, performance data from individual solar panels can be combined with external data, such as electricity grid status, to optimize energy dispatch. This data can be transformed in the Silver layer to ensure uniformity and consistency. In contrast, the Gold layer can provide insights into how changes in grid status affect solar production efficiency.

Another strength of Medallion Architecture in solar energy is its ability to integrate real-time data processing with historical analysis. Real-time data from solar panels can be ingested and processed in the Bronze layer while enriched with historical data in the Silver layer. This enables solar energy companies to monitor current performance against past trends, allowing for rapid adjustments and immediate decision-making based on the system's present state and past behavior patterns.

One key benefit of using Medallion Architecture for solar energy systems is optimizing operational decision-making. Whether determining the optimal time to store energy in batteries or predicting peak solar energy production periods, the data-driven insights derived from the Gold layer enable better decision-making. This can lead to more efficient operations, lower costs, and enhanced performance across the solar energy network.

In conclusion, Medallion Architecture offers a robust, scalable, and flexible approach to handling complex data pipelines in the solar energy industry. Solar companies can unlock the full potential of their data by processing raw data in the Bronze layer, cleaning and enriching it in the Silver layer, and providing actionable insights in the Gold layer. This approach enables predictive maintenance, energy forecasting, and operational optimization, driving greater efficiency and cost-effectiveness in the renewable energy sector.

Conclusion

Medallion Architecture offers a structured approach that enhances data processing efficiency, especially in industries like solar energy, where vast amounts of real-time and historical data need to be processed. By leveraging the Bronze, Silver, and Gold layers, organizations can ensure that data flows seamlessly from raw ingestion to enriched insights, enabling more accurate decision-making.

As demonstrated in the solar energy use case, the Medallion Architecture enables organizations to scale, clean, and optimize data pipelines, unlocking the full potential of predictive analytics and operational efficiency. Companies can improve performance monitoring and forecasting accuracy by integrating real-time data with historical analysis.

The versatility of Medallion Architecture also allows for its application across various industries, not just solar energy. Whether optimizing supply chains, enhancing customer experience, or improving financial modeling, this architecture remains a tried-and-true methodology for delivering actionable insights from large datasets. Its proven principles and adaptability continue to be a relevant and powerful tool for organizations seeking data-driven success.

CHAPTER 24

Kimball Architecture

Kimball Architecture is one of the most widely used frameworks for designing data warehouses. It focuses on creating dimensional models that make data more accessible and understandable for business users. The architecture is built around the concept of star and snowflake schemas, which organize data into facts and dimensions, enabling easy reporting and analysis.

This chapter will delve into the key principles of Kimball Architecture, emphasizing the design of dimensional models, the importance of data marts, and the necessary ETL (Extract, Transform, Load) processes. We will also explore best practices and optimization strategies to maximize the effectiveness of this architecture in data environments.

Overview of Kimball Architecture: Key Principles and Concepts

Kimball Architecture is a robust framework for designing data warehouses, focusing on providing end users easy access to data. The central principle of Kimball is the creation of dimensional models that simplify complex datasets into user-friendly structures. These structures are organized into facts and dimensions, making it easier for business users to analyze and report on data. The ETL (Extract, Transform, Load) process is at the core of this architecture, which ensures that data from various sources is extracted, cleaned, and loaded into a data warehouse for analysis.

A key concept in Kimball Architecture is the star schema, where dimension tables surround a central fact table. The fact table contains quantitative data, while the dimension tables hold descriptive attributes, such as time, customer, and product. This design allows users to query data and obtain valuable insights quickly. Another essential principle is the snowflake schema, which normalizes the dimensions further, splitting

R. L. Steelman, Jr., *Mastering Snowflake DataOps with DataOps.live*, https://doi.org/10.1007/979-8-8688-1754-0_24

them into multiple related tables. Snowflake schemas are typically used for more complex data models that require better storage efficiency and maintainability.

The data mart concept is another essential aspect of Kimball Architecture. Data marts are subsets of data warehouses, often built for specific business needs or departments. These marts ensure that users can access data tailored to their needs without overloading them with unnecessary information. Kimball's approach integrates data marts into the overall data warehouse, ensuring consistency and accuracy across all data models. This makes the architecture flexible and scalable for large organizations with diverse data requirements.

Kimball emphasizes the need for iterative development and continuous refinement to support ongoing changes and growth in the data environment. Data models should be treated as evolving structures, adapting over time to meet the changing needs of the business. This iterative approach allows organizations to start small and expand as necessary, rather than attempting to build an all-encompassing data model from the outset. This flexibility is critical in dynamic business environments where requirements change frequently.

The dimensional modeling methodology is an essential component of the Kimball approach. It focuses on creating intuitive data structures for business users, allowing them to understand and analyze the data quickly. By organizing data into simple, understandable structures, dimensional modeling reduces the complexity that often accompanies relational databases. This makes Kimball Architecture particularly suitable for reporting, ad hoc analysis, and decision-making processes.

Another fundamental principle of the Kimball Architecture is maintaining data consistency across all systems. Ensuring that data remains consistent, accurate, and up-to-date is crucial for generating reliable insights. Kimball advocates for standardizing data definitions, formats, and processes across the organization, which helps eliminate discrepancies and ensures that all stakeholders work from the same information set.

Lastly, performance optimization is a vital aspect of Kimball's design. Since the goal is to provide fast access to large volumes of data, Kimball Architecture employs techniques such as indexing, partitioning, and caching to enhance query performance. These methods ensure that users can perform complex analyses without sacrificing speed, providing both the flexibility and scalability needed for growing data environments.

Designing Dimensional Models: Facts and Dimensions

Designing dimensional models is a cornerstone of the Kimball Architecture, and it begins with defining facts and dimensions. Facts are quantitative, often numeric data that represent business events or transactions. They typically reside in a central table known as the fact table, where they are stored with measurable values. For example, in a sales database, the fact table might include figures like sales revenue, quantity sold, and profit. These values are typically additive and can be summed, averaged, or aggregated for analysis.

On the other hand, dimensions are descriptive attributes that provide context to the facts. Dimension tables store textual or categorical information that helps to interpret the data in the fact table. For instance, in the sales example, dimensions might include time (e.g., year, quarter, month), product (e.g., category, SKU), or customer (e.g., region, name). These attributes allow users to perform slice-and-dice analysis and drill down into the facts based on specific characteristics.

A central feature of dimensional modeling is the star schema, which arranges fact tables in the center and links them to dimension tables through foreign key relationships. This design simplifies queries by enabling users to navigate between facts and dimensions easily. Since dimension tables are typically smaller and more descriptive, this star configuration minimizes the complexity of querying large datasets, enhancing performance for reporting and analysis. The key advantage is that this design makes it intuitive for business users to explore data without complex SQL queries.

For more complex data environments, the snowflake schema comes into play, which is a variation of the star schema. In a snowflake schema, dimension tables are normalized, capable of breaking down into multiple related tables to reduce redundancy. For example, a product dimension might be split into separate tables for product categories, suppliers, and brands. While this design introduces additional tables and relationships, it optimizes storage and ensures better data integrity. However, it also adds complexity to the queries, which might impact performance compared to the more straightforward star schema.

When designing dimensional models, it's vital to define granularity, which refers to the level of detail at which facts are recorded. Granularity influences the fact table's structure and ensures that the data is stored at an appropriate level of precision. For instance, sales data might be stored at the transactional level (each sale recorded individually) or at an aggregated level (monthly sales totals per store). Choosing the correct granularity helps optimize the performance of queries and avoids data duplication or excessive storage use.

Another crucial design decision involves the use of slowly changing dimensions (SCDs). These dimensions capture historical changes to attributes over time. There are several strategies for handling SCDs, depending on whether the business requires tracking historical data changes or simply overwriting old information with new updates. Type 1 SCDs overwrite the old data with the latest, Type 2 SCDs add a new record to capture the change, and Type 3 SCDs store both the old and new values in the same record, with specific fields indicating the changes.

Fact grain is another critical consideration in dimensional design. The grain defines the level of detail in the fact table. For example, if the grain is set to "sales transaction," each row in the fact table will represent an individual transaction. If the grain is set to "daily sales," the fact table will store aggregated sales data by day. The choice of grain impacts query performance, data storage, and the types of insights users can derive from the data.

Once the fact and dimension tables are defined, it's time to focus on indexing and optimization to enhance query performance. Indexing is critical in a dimensional model to speed up searches, especially when working with large datasets. Primary keys uniquely identify records in the dimension tables, while foreign keys establish relationships with the fact table. Additionally, techniques like partitioning, materialized views, and caching can improve performance by reducing the time it takes to retrieve the most commonly queried data.

Surrogate keys are another key concept in dimensional modeling. These are unique identifiers used in the dimension tables, usually in the form of integers, that replace natural keys such as customer IDs or product codes. Surrogate keys simplify the process of managing historical changes to data and improve performance by reducing the complexity of join operations. They ensure that each row in the fact table can be linked to the appropriate record in the dimension table, regardless of changes to natural keys over time.

A clear data dictionary is essential to ensure data consistency and that the dimensional model aligns with business requirements. This dictionary documents all attributes' meaning, format, and constraints in the fact and dimension tables. It serves as a reference for developers and end-users to understand how the data is structured and ensures that different teams use it consistently. A well-maintained data dictionary is significant for large data environments with multiple users.

Finally, business rules should guide the design of both facts and dimensions. These rules help define how data is interpreted and transformed to reflect business requirements. For example, a rule might specify that all sales transactions should exclude returns or that the product dimension should categorize products into predefined groups. Ensuring that the dimensional model aligns with the business's needs guarantees that the insights derived from the data will be relevant and actionable. By aligning the model with these rules, companies can increase the effectiveness of data-driven decision-making.

By carefully designing dimensional models with well-defined facts and dimensions, businesses can build an analytical framework that is intuitive, flexible, and optimized for performance. The goal is to make it as easy as possible for end-users to interact with the data and gain valuable insights without needing deep technical expertise. Whether working with a star or snowflake schema, managing slow-changing dimensions, or optimizing query performance, these design principles ensure that the dimensional model efficiently supports business intelligence goals.

Building Data Marts: Organizing Data for Business Use

Building data marts is an essential step in organizing data for business use, as it helps centralize and structure data to make it easily accessible for specific business functions. A data mart is a data warehouse subset typically focused on a particular department, business area, or functional unit. Unlike enterprise-wide data warehouses that aggregate data for the entire organization, data marts are designed to meet the specific needs of individual business users or teams, such as marketing, finance, or sales.

Building a data mart starts by identifying the business requirements. Understanding the key performance indicators (KPIs), metrics, and reporting needs of the target business unit is critical. These insights will guide the selection of relevant data sources, which can come from operational systems, external datasets, or enterprise data warehouses. This approach ensures that the data mart contains only the necessary and most relevant information for that department's decision-making process, avoiding unnecessary complexity.

A crucial step in building a data mart is the data integration process. This involves pulling together data from different systems into a cohesive structure that is consistent, clean, and ready for analysis. Data integration may include data extraction, transformation, and loading (ETL), which prepares the data by cleaning, transforming, and loading it into the data mart. During this process, it's essential to ensure data quality and consistency so that users can rely on the insights provided by the data mart.

After integration, the next step is defining the schema and structure of the data mart. Typically, this involves organizing the data in a star schema or snowflake schema, which optimizes the ability to perform queries and analyses. In a star schema, a central fact table is surrounded by dimension tables, while a snowflake schema normalizes the dimension tables into multiple related tables. The schema design should reflect how business users interact with the data and the types of analyses they need to perform.

Data security and governance are essential considerations when building a data mart. Since data marts typically contain sensitive business information, such as customer or financial data, it's important to implement proper access controls, encryption, and audit trails. Data governance ensures that the data is consistent, accurate, and compliant with regulations. Well-defined roles and permissions should be set up to control who can view, modify, or manage the data within the data mart.

Once the data mart is built, it's important to maintain and update it regularly. This includes refreshing the data as new information becomes available and adjusting the schema as business needs evolve. Regularly monitoring the data mart's performance is also necessary to ensure that query times remain fast and that it meets user needs. As business priorities shift, the data mart's structure and data sources may need to be modified to reflect new business requirements.

Finally, data marts help empower business users by giving them easy access to relevant data without navigating complex enterprise-wide systems. By providing a more focused, tailored view of the data, data marts enable quicker decision-making, more efficient reporting, and deeper insights into specific business areas. This level of customization allows users to analyze and interpret data in ways that are directly aligned with their objectives, leading to more informed and effective business strategies.

ETL Processes in Kimball Architecture

In the Kimball architecture, the ETL (Extract, Transform, Load) process plays a pivotal role in data integration, ensuring that data is correctly extracted from source systems, transformed into the desired format, and loaded into the data warehouse or data mart. The ETL process begins with data extraction, which involves pulling data from various heterogeneous sources such as transactional databases, flat files, or external APIs. This step ensures that all relevant data is gathered for analysis, though it requires careful attention to source system compatibility and extraction frequency.

Once the data is extracted, the next stage is data transformation. This involves cleaning, filtering, and converting the raw data into a format that is optimized for analysis. Transformation steps include removing duplicates, handling missing values, and applying business rules to standardize the data. Additionally, data may be aggregated or calculated at this stage to meet reporting requirements. The transformation step is critical for ensuring data consistency, quality, and accuracy, enabling users to make reliable business decisions.

After transformation, the loading process occurs, where the data is moved into the data warehouse or data mart. Kimball's approach typically loads the data into a star or snowflake schema. These schemas are designed to make querying faster and easier by structuring data in a way that is intuitive to business users. During this phase, the data is carefully mapped to the schema, and indices are often created to improve query performance. The loading process also includes routine updates to ensure that new data is captured and processed continuously.

In Kimball architecture, the ETL process is often designed with incremental loads to optimize performance and minimize processing time. Instead of reloading the entire dataset, only new or changed data is extracted and loaded, reducing the time and resources required for ETL operations. This is particularly important in large-scale data environments where full data reloads could lead to performance bottlenecks. Incremental ETL also ensures that the data warehouse is regularly updated with the most current information available.

ETL processes are crucial for data consistency and data quality management in the Kimball architecture. Since the ETL process transforms and cleans data, robust validation and error-handling mechanisms are essential. Automated tests and validation checks can detect and correct issues during the ETL process, ensuring that data loaded into the warehouse is accurate, complete, and trustworthy. By maintaining high data quality, ETL ensures that business users have reliable data for decision-making.

Parallel processing and automation are frequently employed to optimize ETL performance. Parallel processing allows multiple ETL jobs to run concurrently, significantly speeding up the extraction, transformation, and loading processes. Conversely, automation ensures that ETL workflows are executed on a consistent schedule, reducing the need for manual intervention and improving overall efficiency. Both approaches are essential for maintaining a high-performing and scalable ETL pipeline within the Kimball architecture, especially as data volumes grow over time.

Kimball Architecture Best Practices and Optimization Strategies

Kimball Architecture provides a robust framework for building efficient data warehouses that cater to business intelligence needs. Best practices in Kimball architecture focus on ensuring data is accessible, accurate, and useful for decision-making. One key best practice is to design with the end user in mind, ensuring that the data is structured in a way that aligns with business processes and reporting needs. This means utilizing star or snowflake schemas to simplify querying and ensure data is easily interpretable for non-technical users.

Another important practice is ensuring that ETL processes are optimized for performance and data accuracy. Data should be cleansed and transformed before being loaded into the warehouse, ensuring it is ready for use immediately. Incremental loading should be considered wherever possible to reduce processing time and make the ETL process more efficient. This helps keep the data up-to-date without overburdening the system with unnecessary reloads.

Performance optimization is also crucial in Kimball architecture. Indexing and partitioning data are essential strategies for improving query performance. Indexing helps speed up data retrieval, while partitioning divides large datasets into smaller, more manageable pieces, reducing the time it takes to query the data. Additionally, having a good strategy for data aggregation helps provide faster access to summary-level data, which is often more useful to business users than detailed transactional data.

Modular design is recommended to improve scalability and maintainability. Kimball architecture emphasizes creating scalable data marts and ensuring they align with specific business domains or processes. This allows organizations to grow their data infrastructure over time without disrupting existing workflows. Each data mart should focus on a particular subject area, and the architecture should enable easy integration of new data sources as the business grows.

Another best practice is to prioritize data governance and security throughout the architecture. Properly securing the data and ensuring its integrity is essential for building trust with business users and preventing unauthorized access. Implementing data lineage tracking, audit trails, and role-based access controls helps maintain transparency and accountability in the data warehouse.

Finally, Kimball architecture thrives in an agile environment. The ability to iterate and make adjustments as business needs evolve is vital for long-term success. Data requirements may change, and the architecture must be flexible enough to accommodate these changes without requiring a complete redesign. An agile approach ensures that data can be continuously improved, adjusted, and optimized based on user feedback and business needs.

Conclusion

Kimball architecture continues to be a cornerstone of data warehousing due to its clear structure and focus on usability. The principles laid out in this approach help organizations create data systems that are both scalable and efficient, allowing business users to access and derive insights from their data efficiently. By emphasizing dimensional modeling, streamlined ETL processes, and precise data marts, Kimball ensures that the data warehouse is both practical and adaptable to evolving business needs.

The architecture's modular design facilitates growth, ensuring businesses can expand their data infrastructure without disrupting existing processes. This flexibility, combined with best practices for performance optimization and governance, makes Kimball a durable model for data warehousing. Proper implementation can yield fast, accessible, and reliable data supporting business decision-making.

As businesses continue to collect vast amounts of data, the Kimball architecture remains a tried-and-true method for ensuring data is organized, accurate, and easily accessible. By following the principles outlined in this chapter, organizations can maintain a data warehouse that scales with their needs, promotes strong data governance, and enables efficient, actionable analytics.

CHAPTER 25

DataVault 2.0 Architecture

DataVault 2.0 is a modern, scalable approach to data warehousing designed to handle the complexities of large-scale data environments. It provides a flexible framework for managing rapidly changing data and supporting business agility, making it an ideal solution for organizations navigating significant data challenges. Unlike traditional methodologies, DataVault 2.0 focuses on modeling and integrating disparate data sources with a focus on traceability, auditability, and scalability.

In this chapter, we will explore the foundational principles and components of DataVault 2.0, providing insights into its advantages over other architectures. We'll also discuss its real-world applications, best practices, and how it integrates with modern DataOps methodologies to enhance operational efficiency and improve decision-making processes.

Introduction to DataVault 2.0 Architecture

DataVault 2.0 architecture is designed to address the complexities of modern data environments. It offers a comprehensive solution for managing large volumes of data from disparate sources while ensuring the traceability, auditability, and scalability required for business intelligence. By focusing on agile data modeling, DataVault 2.0 enables organizations to integrate historical and real-time data, ensuring they can quickly adapt to changing business needs and evolving data requirements. This architecture ensures business teams can access accurate and timely information to drive decision-making.

The core concept behind DataVault 2.0 is to establish a robust, flexible, and scalable framework for managing data. It emphasizes the importance of capturing the raw, unmodified data securely and in an organized manner, making it possible to trace, audit, and manage the data lineage. Unlike traditional star or snowflake schemas, DataVault 2.0

R. L. Steelman, Jr., *Mastering Snowflake DataOps with DataOps.live*, https://doi.org/10.1007/979-8-8688-1754-0_25

structures data into three key components: hubs, links, and satellites. These components provide a way to track and manage changes in data sources over time, ensuring that the data remains consistent and accessible.

Hubs represent an organization's core business entities, such as customers or products, and are identified by unique keys. Links define the relationships between these business entities, capturing the associations or transactions that occur between them. Satellites store additional attributes and historical information related to the hubs and links, allowing for data preservation in its raw form, including changes over time. These components form a flexible, dynamic, and scalable framework for storing and managing data.

DataVault 2.0 encourages an iterative approach to data modeling, where new data sources can be incorporated as needed. This modular design ensures that the data warehouse can scale as business needs grow, allowing organizations to easily add new hubs, links, and satellites without disrupting existing processes. As a result, DataVault 2.0 is well-suited to handle complex data environments that require flexibility and adaptability in data management.

One of DataVault 2.0's key benefits is its ability to provide a single source of truth for all organizational data. Capturing raw data from various sources and organizing it to be easily traced and audited helps ensure data integrity and accuracy. This approach is particularly beneficial in industries where data is constantly evolving, such as finance, healthcare, and e-commerce, where data lineage and transparency are essential for regulatory compliance and business decision-making.

Moreover, DataVault 2.0 supports real-time data integration, which is critical for modern data-driven businesses. With the rise of IoT, machine learning, and real-time analytics, companies need the ability to process and integrate data from various sources quickly and efficiently. DataVault 2.0 enables this by providing a flexible and scalable platform for integrating and managing data from batch and real-time sources. This makes it easier for businesses to adapt to changing market conditions and stay competitive in their respective industries.

Another essential feature of DataVault 2.0 is its focus on automation. By using automation tools, organizations can streamline data integration processes, reducing the time and effort required to manage and maintain their data architecture. This also minimizes the risk of human error, ensuring that data is consistently and accurately integrated into the system. Automation enables organizations to focus more on deriving insights from the data rather than spending time on manual data management tasks.

DataVault 2.0 also emphasizes the importance of data governance and security. By capturing data in its raw form and tracking changes over time, organizations can ensure that they have complete visibility into their data and its lineage. This helps with compliance and auditing and ensures that sensitive data is handled securely. The architecture allows for granular control over data access and security, ensuring that only authorized users can access specific data sets.

Finally, DataVault 2.0 is a future-proof architecture built to handle the growing complexities of modern data environments. As organizations continue to generate increasing volumes of data from diverse sources, the ability to scale and adapt is crucial. DataVault 2.0 provides a flexible and scalable platform that can accommodate this growth, ensuring that businesses can continue to derive value from their data for years to come. Its focus on agility, scalability, and automation makes it an ideal choice for organizations looking to modernize their data infrastructure.

Key Principles of DataVault 2.0

DataVault 2.0 is centered around several key principles that help it remain flexible, scalable, and capable of handling modern organizations' complex, evolving data environments. These principles provide the foundation for data architecture that can quickly adapt to changes and scale over time while ensuring the accuracy and integrity of the data. The central tenets of DataVault 2.0 include agility, traceability, scalability, modularity, and automation.

The principle of agility emphasizes the need for the data model to adapt to changes quickly. Unlike traditional approaches that may require significant reworking of the data model when new business requirements emerge, DataVault 2.0 allows organizations to integrate new data sources, business requirements, or updates to existing data with minimal effort. This flexibility ensures that the architecture remains relevant as business needs evolve, without the need for significant disruptions or overhauls.

Traceability is another core principle of DataVault 2.0. The architecture ensures that every piece of data is traceable from its source to its usage in the final data mart or report. By capturing raw, unmodified data from various sources, DataVault 2.0 allows complete visibility into the data lineage, which is critical for auditing, compliance, and ensuring data quality. This traceability ensures that organizations can confidently rely on their data and demonstrate its integrity to stakeholders and regulatory bodies.

Scalability is a fundamental requirement for DataVault 2.0, as modern data environments often deal with vast amounts of data. DataVault 2.0's architecture is designed to scale easily, allowing organizations to grow their data environments without needing significant rework. It can handle batch and real-time data ingestion, ensuring businesses can integrate diverse data sources and maintain performance as data volumes increase. Whether an organization is dealing with terabytes or petabytes of data, DataVault 2.0 provides the infrastructure to scale as needed.

Modularity is another essential principle in DataVault 2.0. The architecture uses independent components, including hubs, links, and satellites, that work together to form a cohesive data structure. This modular design allows organizations to add, remove, or modify components of the data model without affecting the entire system. It also means that teams can work on different parts of the data model concurrently, improving collaboration and reducing the time required to integrate new data sources or business requirements.

Automation significantly affects DataVault 2.0's ability to streamline data integration processes. By automating key parts of the data pipeline, including data ingestion, transformation, and quality checks, DataVault 2.0 minimizes manual intervention and reduces the risk of errors. Automation also accelerates the process of updating data models, making it easier for organizations to keep their data infrastructure up-to-date and aligned with evolving business needs. This efficiency allows data teams to focus on deriving insights rather than managing complex manual processes.

Another key principle of DataVault 2.0 is its focus on historical data preservation. In contrast to other architectures that may discard older data as systems evolve, DataVault 2.0 ensures that historical data is retained in its raw form, including all changes made to it over time. This approach is important for organizations that must track data changes over time for compliance, auditing, or business analysis purposes. By storing historical data in satellites, DataVault 2.0 ensures that all versions of the data are preserved for future reference.

The DataVault 2.0 architecture also promotes a business-driven approach. By organizing data into hubs, links, and satellites that represent key business entities, DataVault 2.0 allows data to be aligned with the organization's business model. This makes it easier for business users to understand and leverage the data, as it is structured to mirror how the organization operates. This alignment between business needs and the data model ensures that data is accessible and valuable to organizational stakeholders.

One of DataVault 2.0's overarching principles is to provide a flexible, future-proof solution. Given the ever-changing landscape of data technologies and business requirements, DataVault 2.0 is built to accommodate new data types, sources, and technologies without requiring a complete redesign. This ensures that organizations can continue to adapt to new trends, like machine learning or IoT, while still maintaining a cohesive data architecture. The system's modularity and scalability enable organizations to meet both current and future data challenges.

DataVault 2.0 also emphasizes the importance of data governance. The architecture's traceability and history-preserving features make it easier for organizations to maintain compliance with data privacy laws and other regulations. By keeping track of the raw data and all changes made to it, DataVault 2.0 ensures that data governance requirements are met without adding unnecessary complexity. This focus on governance also facilitates data security, as access can be controlled and monitored to ensure that only authorized individuals can interact with sensitive information.

Finally, DataVault 2.0 supports batch and real-time data processing, making it a versatile choice for organizations with diverse data needs. This flexibility allows businesses to process large volumes of batch data and smaller, real-time streams, enabling them to respond more quickly to new data and business opportunities. The combination of historical data preservation, real-time processing, and automation ensures that DataVault 2.0 can handle various use cases and data scenarios, making it a highly adaptable and practical architecture for modern data environments.

In summary, the principles of agility, traceability, scalability, modularity, automation, historical preservation, business alignment, and data governance come together to form the foundation of DataVault 2.0. These principles ensure that organizations can manage their data flexibly, efficiently, and future-proof, enabling them to stay competitive and compliant while meeting the demands of modern business and technology landscapes.

Core Components: Hubs, Links, and Satellites

In DataVault 2.0, the core components—hubs, links, and satellites—form the backbone of the architecture, each serving a distinct role. Hubs capture unique business keys, ensuring that each entity, such as a customer or product, is represented with a consistent identifier. These business keys are central to the DataVault model, allowing for seamless data integration across different systems and data sources. Hubs act as the core point of reference for all other components within the architecture.

Links establish relationships between different hubs. They are designed to represent the associations or connections between entities. For instance, a link might describe the relationship between a customer and an order, allowing for dynamic and flexible querying across business domains. Links maintain the integrity of the relationships between business keys, and their inclusion ensures that DataVault 2.0 can support complex, interconnected data structures while preserving traceability and auditability.

Satellites store descriptive or contextual data about the business keys defined in the hubs. These elements contain additional attributes that provide a deeper understanding of the business entities, such as customer demographics or product specifications. Satellites can evolve, capturing historical changes and ensuring the data is comprehensive and up-to-date. By keeping the descriptive data separate from the keys and relationships, satellites provide flexibility and scalability without compromising performance or data quality.

Together, hubs, links, and satellites create a modular and adaptable architecture that supports agility, traceability, and historical preservation. Each component is independent yet interconnected, enabling organizations to scale their data environments and easily integrate new data sources. The separation of concerns allows DataVault 2.0 to maintain high performance while still accommodating complex and evolving business requirements. This architecture also simplifies the process of adding new data models, making it a robust and scalable solution for modern data environments.

DataVault Hubs

In DataVault 2.0 architecture, hubs are foundational elements that capture and store unique business keys for each core entity within the system. These business keys are identifiers that ensure data integrity and traceability across multiple data systems. For example, a hub might represent an entity such as a customer, product, or order, with the business key serving as the unique identifier for each of these entities. Hubs allow for data integration from various sources without the need for complex transformations or data reshaping, preserving the raw, unaltered value of the key.

Each hub contains the business key, a unique identifier that acts as the core reference point for all data related to that entity. This business key is used to link related data from other hubs, links, and satellites in the DataVault. The structure of the hub ensures that the same business key can be reused across various data sources, simplifying the process of consolidating data from multiple systems. By acting as the central reference point, hubs ensure that entities are consistently identified, even when sourced from disparate systems.

The key principle behind the hub is its independence from other components. This independence allows hubs to be lightweight and efficient, avoiding unnecessary dependencies that might complicate future changes or scaling. A hub is designed to be immutable, meaning that once a business key is captured, it is never modified. Instead, if there is a change in the attributes related to that business key, it is captured in a satellite, leaving the hub unchanged. This design ensures historical accuracy and enables a clear, unbroken reference to the original data.

Regarding architecture, hubs are typically small and contain minimal metadata beyond the business key itself. This streamlined design allows for quick identification and access to the data without the overhead of unnecessary details. For instance, a hub might only contain the business key, load date, and an identifier for the source system from which the data originated. The hub's minimalism ensures it remains performant, even as data volumes increase.

Hubs serve as the primary building blocks for creating a scalable and flexible data model. Because the hub stores only the business key, it allows data professionals to add new data sources or change source systems without disrupting the core structure. When a new system is integrated into the architecture, only a new hub is required, ensuring that the DataVault can grow organically as the business evolves. This modularity also makes it easier to replace or update systems without having to perform significant refactoring.

A critical aspect of DataVault hubs is their ability to handle late-arriving data. DataVault is designed to accept data even after it has been processed by other systems, making it easier to manage discrepancies and delays in data ingestion. This flexibility ensures that all data is captured, even if it arrives out of sequence. The hub design supports this by enabling data to be inserted with its timestamp and source metadata, ensuring that the data can still be accurately traced back to the original business key and integrated into the overall model.

Hubs also ensure that data remains consistent and auditable across different systems. DataVault simplifies data reconciliation from various sources using a single, consistent business key. This consistency is crucial in environments where data is ingested from multiple systems that may have varying definitions or representations of the same entities. By relying on a central business key, hubs provide a unified view of data that is easier to track and manage, ensuring that data remains accurate and transparent.

Another important feature of hubs is their support for slowly changing dimensions (SCD). Since the hub only stores the business key and its associated metadata, changes in the attributes or relationships of the business entity are captured in satellites or links, not in the hub itself. This separation ensures that historical data is preserved, while the hub remains a stable, unchanging entity. This approach aligns with the principle of data versioning, allowing businesses to track and maintain multiple versions of the same entity over time.

Hubs also facilitate the easy creation of relationship mappings between different data models. Since hubs anchor other components like links and satellites, they enable a flexible and dynamic way to represent relationships between business entities. For example, a link might connect two hubs representing a customer and an order, while a satellite might capture the details of a customer's interactions or orders over time. This flexible relationship modeling is one of the reasons why DataVault is well-suited for dynamic and complex data environments.

The hub's role is central to the overall DataVault methodology, acting as the source of truth for each business key. This centralization is key to DataVault's scalability and flexibility, allowing for easy data model expansion and modification. When new data sources or new business rules are introduced, they can be incorporated without disrupting the existing structure. The hub is designed to be adaptable to the needs of the business, ensuring that the data model evolves alongside organizational changes.

In summary, the DataVault hub is a critical component that anchors all other data structures within the architecture. Its primary function is to capture and store the business key, ensuring data consistency, traceability, and scalability. By separating the key from descriptive attributes, DataVault ensures that the data remains flexible and can evolve as business requirements change. The hub's simplicity and ability to handle late-arriving data and support slowly changing dimensions make it an essential part of the DataVault 2.0 model.

DataVault Links

In the DataVault 2.0 architecture, links are crucial in connecting different hubs and representing the relationships between various business entities. Links are designed to capture the many-to-many relationships that exist between hubs, ensuring that data is properly connected and can be traced back to its source. Unlike hubs, which contain only the business key, links include both the business keys of the related entities and the metadata necessary to define the relationship between them. Links can be used to model various types of connections, from customer-product relationships to employee-project assignments.

For example, a link might be used to represent the relationship between customers and orders in an e-commerce environment. The link table would contain the business key for the customer (from the Customer hub) and the business key for the order (from the Order hub). This establishes a direct connection between the customer and their purchases, making it easier to analyze order history, customer behavior, and sales patterns. In addition to the two business keys, the link could include the date the order was placed, the order status, and the sales region.

Another example of a link is the relationship between employees and the projects they are assigned to. The Employee and Project hubs would be connected through a link table that includes the employee's business key and the project's business key. Additional data elements in this link might include the employee's role in the project, the project start and end dates, and the project status. By modeling this relationship in a link, businesses can track employee workload, project progress, and resource allocation over time.

In some cases, links can be used to represent complex relationships that involve multiple hubs. For instance, in a healthcare environment, a link might be used to connect patients, doctors, and treatments. The Patient hub, Doctor hub, and Treatment hub would all be connected through a single link table, which would store the business keys of each entity. Additional data elements in this link could include the date of the treatment, the type of procedure performed, and the location of the treatment. This structure would enable healthcare providers to analyze patient care across different doctors and treatment types.

Links can also represent time-dependent relationships. For example, in a retail setting, a link might track the changing relationship between a customer and a particular product over time. The link table could include the customer's business key, the product's business key, and the date the customer interacted with the product, such as a purchase or a return. This enables tracking customer preferences and buying patterns, allowing for more targeted marketing and personalized recommendations.

A critical feature of link tables is that they can store multiple relationships between the same pair of hubs. For instance, a customer may have a purchase relationship and a service request relationship with the same product. In this case, the link table would include both relationships, each identified by a unique business key combination and with different timestamps or statuses to distinguish between them. This ability to model multiple relationships in a single table makes links highly flexible for various use cases.

Link tables in DataVault also support the handling of slowly changing relationships. Suppose a relationship between two business entities evolves over time, such as a change in the customer's loyalty program status or a change in the project's scope. In that case, the link table can accommodate this change without disrupting the data integrity of the underlying hubs. This is achieved by including time-stamping information in the link, allowing for tracking of historical changes and ensuring that relationships are accurately represented over time.

A link table might connect customers, sales representatives, and products in a sales environment. The link would contain the business keys for each entity and also store the transaction date, the sales rep who handled the sale, and the quantity of the product purchased. This enables sales managers to analyze performance by representative, track sales trends for different products, and examine how customer behavior evolves over time in response to various sales strategies.

In addition to business keys, link tables can also store descriptive metadata to define the relationship's nature further. For example, in a manufacturing context, a link might be used to represent the relationship between products and suppliers. The link table could include the supplier's business key, the product's business key, and metadata such as delivery dates, lead times, and contract terms. This additional data allows for in-depth analysis of supply chain efficiency, product availability, and vendor performance.

Link tables can also be leveraged for analytics purposes, such as calculating key performance indicators (KPIs) or conducting trend analysis. Organizations can gain deeper insights into their operations by linking different business entities together. For example, a link table that connects employees to training programs might allow an organization to analyze the correlation between training attendance and employee performance or retention rates.

Finally, links are integral to maintaining the flexibility and scalability of the DataVault 2.0 architecture. Since links are designed to connect hubs flexibly and modularly, adding new relationships between business entities does not require significant restructuring. For example, suppose a new relationship is identified between a customer and a service provider. In that case, a new link table can be created, or an existing link table can be modified to include the new relationship. This makes DataVault a powerful solution for dynamic and evolving data environments.

In conclusion, links in the DataVault 2.0 architecture are essential for modeling relationships between business entities. They enable organizations to connect disparate data sources, track changes over time, and better understand their operations. By including business keys and relevant metadata, links offer a powerful way to represent complex relationships and support analytics, making them a cornerstone of the DataVault methodology.

DataVault Satellites

In the DataVault 2.0 architecture, satellites are a critical component that captures historical changes to the descriptive attributes of business entities. While hubs contain the business keys and links store relationships between those keys, satellites store the attributes, or the "context" of the business entities, such as customer details or product descriptions. The purpose of a satellite is to store the changes over time, maintaining an accurate and historical record of how business data evolves. This separation of business keys from descriptive data allows for flexibility, scalability, and improved performance in data warehousing.

Each satellite is tied to a specific hub or link but contains multiple pieces of information. For example, in the case of a Customer hub, the satellite might store the customer's address, email, and phone number. If the customer updates their email address, the satellite would track this change and associate it with the same business key from the Customer hub, along with a timestamp to preserve historical context. By doing so, satellites ensure that data is never lost, and the history of changes can be easily queried.

One of the most essential characteristics of a satellite is its ability to handle slowly changing dimensions. A common challenge in data warehousing is managing data that evolves over time, such as customer demographics or product specification changes. In a traditional schema, these changes might result in overwriting the previous values. However, DataVault satellites keep track of all historical changes, so every data version is preserved and available for analysis. This enables businesses to analyze trends over time and build a comprehensive view of the data's evolution.

Satellites in DataVault 2.0 architecture are typically designed to be as granular as possible, capturing every change to an entity's attributes. This aligns with the DataVault methodology, emphasizing the importance of having detailed, auditable, and accurate historical data. In practice, this means that a single hub might have multiple satellites connected to it, each capturing a different aspect of the entity's characteristics. For example, a hub representing products might have satellites for product descriptions, pricing history, inventory levels, and promotional details.

The design of satellites ensures that they are both flexible and optimized for performance. Since satellites are designed only to store changes, they avoid the issue of storing redundant or unnecessary information. This approach reduces the overall storage footprint and ensures that only the relevant and changed data is stored over time. This methodology is also advantageous for querying, as it allows users to retrieve only the most recent or specific versions of data when needed, without having to scan through a massive dataset that includes outdated information.

Satellites are often designed to handle different types of data updates efficiently to optimize performance further. For example, some satellites might be designed to store data in a "full history" mode, where every change is stored in detail. In contrast, others may focus on "delta" updates, only capturing the modifications made to an entity's attributes. The choice of approach depends on the business's specific requirements, and some environments may benefit from a hybrid approach that combines both methods.

The granularity of data captured in a satellite can vary, but keeping the data as atomic as possible is generally advisable. For example, in a sales environment, instead of having a single satellite that tracks all customer data, separate satellites might be created to track customer contact details, sales interactions, and loyalty program status. This enables more precise querying, and it helps reduce the risk of overwriting or losing critical historical information.

An important benefit of the satellite layer in DataVault 2.0 is that it allows for a highly modular design. Since each satellite is tied to a specific hub or link and only captures certain aspects of data, they can be created and modified independently of other parts of the architecture. This modularity supports both agile development and rapid scalability. As new business needs emerge, new satellites can be added without impacting the core structure of the DataVault, making it easy to adapt to evolving requirements.

Another essential feature of satellites is their role in ensuring data quality and consistency. By capturing all changes to an entity's attributes, satellites help to maintain an accurate historical record of the business entity, enabling users to detect errors or inconsistencies that may arise over time. For example, suppose there is a sudden change in the pricing of a product. In that case, the satellite will record this change, allowing for immediate identification of potential issues and discrepancies in pricing history.

One challenge organizations might face when working with satellites is the potential volume of data. Since each change to an entity's attributes is captured separately, organizations must ensure that their infrastructure can handle the high data throughput and storage requirements that come with maintaining historical data. Fortunately,

DataVault's design allows for incremental data loading, meaning that only new changes need to be captured, thus mitigating some of the performance overhead that could arise from storing large amounts of historical data.

Despite the increased volume of data satellites may create, their ability to store detailed historical information is crucial for several business processes. For instance, when performing trend analysis or building predictive models, it is essential to have access to a complete and detailed history of an entity's attributes. By using satellites to preserve this historical data, organizations can gain deeper insights into business trends, customer behavior, and market fluctuations.

A well-designed satellite also improves data governance and auditability. Since each change to an entity's attributes is stored separately with a timestamp and a unique record identifier, tracking the data's provenance and ensuring compliance with regulatory standards is easier. For example, data lineage is crucial for maintaining legal and regulatory compliance in industries like healthcare or finance. By using satellites, organizations can demonstrate the accuracy and integrity of their data, which is essential for audits and reporting.

In summary, satellites are integral to the DataVault 2.0 architecture, as they store business entities' historical context and descriptive attributes. They provide flexibility and scalability and ensure that all changes to data are captured and available for analysis. By enabling organizations to track slowly changing dimensions, optimize data storage, and maintain historical records of changes, satellites enhance the overall functionality of the DataVault and support advanced analytics and reporting capabilities.

Benefits of DataVault 2.0 for Modern Data Environments

DataVault 2.0 offers significant advantages for modern data environments, particularly in handling large and complex data ecosystems. One of the primary benefits is its scalability, allowing organizations to manage vast amounts of data while maintaining performance. The flexible design of DataVault 2.0 enables businesses to expand their data models without disrupting existing systems, making it ideal for continuously evolving environments.

The architecture is designed to support agility and quick adaptation to changing business requirements. Traditional data warehousing models could be time-consuming and costly for adapting to new data sources or evolving business needs. However, DataVault 2.0's modular approach makes introducing new data sources and business rules easier, reducing the lead time required to make necessary changes.

DataVault 2.0 also excels in handling data from disparate sources. Its ability to ingest and integrate data from various systems without significant rework is a critical advantage for modern enterprises that deal with diverse data sources. This flexibility allows businesses to build a unified, accurate view of their data, regardless of its origin or format, enabling more comprehensive analytics and insights.

Another benefit of DataVault 2.0 is its robustness in dealing with historical data. By capturing and maintaining detailed records of changes over time, the architecture provides a complete audit trail, making it easier for organizations to ensure data integrity and compliance with regulatory standards. This historical perspective also enables businesses to perform advanced trend analysis and predictive modeling.

Moreover, DataVault 2.0 encourages data governance and security. Its structure allows organizations to manage sensitive data more effectively, ensuring proper access control and monitoring. This is especially important for industries with strict data privacy requirements, such as healthcare and finance, where safeguarding data is critical to avoiding compliance violations.

The approach promotes collaboration among business and technical teams. Separating business keys, relationships, and descriptive data into distinct components (hubs, links, and satellites) allows data engineers, analysts, and business users to work more independently while maintaining a coherent data model. This separation fosters greater transparency and enables teams to work efficiently without overlapping responsibilities.

DataVault 2.0 also improves data quality and consistency across the enterprise. Because the system tracks changes over time and integrates data from various sources, it helps organizations identify discrepancies early in the data lifecycle. This results in cleaner, more reliable data, which is essential for accurate reporting and decision-making.

One of the more practical benefits is its alignment with modern cloud-based infrastructures. As more organizations shift to cloud platforms, DataVault 2.0's ability to integrate with cloud technologies, such as Snowflake and Google BigQuery, becomes increasingly important. Cloud-native tools that support DataVault principles allow businesses to leverage their data more efficiently while benefiting from scalability and cost-effectiveness.

Finally, DataVault 2.0 is designed to support advanced analytics and machine learning. The comprehensive and historical nature of the data captured in DataVault makes it ideal for building machine learning models and conducting advanced analytics. By having access to clean, governed, and well-structured data, businesses can gain deeper insights and build more accurate predictive models, which is crucial in today's data-driven environment.

In conclusion, DataVault 2.0's flexibility, scalability, and ability to handle complex data environments make it a suitable choice for modern enterprises. Whether integrating diverse data sources, managing historical records, or ensuring regulatory compliance, DataVault 2.0 offers the features necessary to drive successful data initiatives in dynamic and evolving business landscapes.

Implementing DataVault 2.0 with DataOps.live

Implementing DataVault 2.0 with DataOps.live requires a well-planned approach to effectively leverage the platform's capabilities while aligning with DataVault's principles. The first step is establishing a structured data pipeline that captures and integrates data from various sources into DataOps.live. This allows users to efficiently organize and store raw data within hubs and links, adhering to the architecture's design of separating business keys, relationships, and descriptive data.

A critical part of this implementation is ensuring that the data in the hubs, links, and satellites is governed correctly. DataOps.live offers robust governance tools, such as version control and lineage tracking, which are essential to manage DataVault's historical data. By using DataOps.live's versioning and automated data validation features, organizations can ensure that data integrity is maintained, providing a clear audit trail across the data lifecycle.

Another important factor in implementing DataVault 2.0 with DataOps.live is automation. The platform supports the automation of the ETL processes, a core component of DataVault 2.0. Automating the extraction, transformation, and loading of data ensures that data is consistently ingested and transformed as it enters each stage of the DataVault model, minimizing manual intervention and increasing reliability. With this automation, DataOps.live can streamline the process of populating the hubs, links, and satellites with high-quality data.

DataOps.live also plays a vital role in optimizing the performance of a DataVault 2.0 implementation. The platform can handle large-scale data environments by utilizing cloud-native capabilities such as elastic scalability and parallel processing, ensuring fast data processing. DataOps.live's built-in support for cloud technologies like Snowflake and BigQuery allows businesses to quickly scale their DataVault infrastructure without sacrificing performance.

In addition to performance optimization, DataOps.live helps improve collaboration between business and technical teams. As DataVault 2.0 encourages the separation of data into distinct layers, DataOps.live's collaborative features enable teams to work on different components independently, reducing bottlenecks. Business analysts can access and explore the data through self-service capabilities, while technical teams focus on ensuring the smooth operation of the underlying data pipelines.

One significant advantage of integrating DataVault 2.0 with DataOps.live is its ability to support continuous data integration. DataOps.live facilitates a continuous delivery pipeline for data models, ensuring that the data architecture evolves in tandem with business needs. This allows businesses to quickly adapt to changes in data sources or business requirements, enabling a more agile and responsive data environment.

Lastly, DataOps.live provides an extensive set of monitoring and alerting features crucial for the ongoing health of DataVault 2.0 implementations. These tools enable teams to track data quality, performance, and any potential errors in real time. With detailed metrics and alerts, organizations can proactively address issues, ensuring minimal disruptions to the data flow and maintaining the integrity of the data stored within the DataVault system.

Best Practices for DataVault 2.0 Modeling

Best practices for DataVault 2.0 modeling are critical for ensuring data integrity, scalability, and agility in a modern data environment. The first best practice is to focus on creating clear and concise hubs. A well-defined hub captures unique business keys without including descriptive data, allowing for data consistency across the enterprise. Keeping the hub tables lean and focused on business keys ensures that the architecture remains flexible and adaptable to changes in source systems.

The next best practice revolves around the proper use of links. Links are used to capture the relationships between business keys, and it's essential that each link table clearly defines these relationships. Properly defining link tables ensures they can support various reporting and analytical needs. It's essential to avoid unnecessary complexity in link tables, which can be achieved by keeping them as straightforward as possible while ensuring they meet the business requirements.

Best practices for satellites emphasize the importance of maintaining a history of descriptive data without introducing data duplication. This is done by creating time-stamped records that track changes over time. Satellites should be designed to hold data that changes over time, such as customer details or transaction information. Using a unique surrogate key for each satellite record helps prevent data inconsistencies and allows for accurate time-based analysis.

One key best practice in DataVault 2.0 is ensuring that each architecture component, including hubs, links, and satellites, is designed with scalability in mind. As organizations grow and collect more data, their data vaults must scale efficiently. Using cloud-based storage and distributed processing can help accommodate growing data volumes and ensure optimal performance even as the data set increases.

Another important practice is adopting an agile approach to DataVault 2.0 development. Given the fast-paced nature of modern business, flexibility is paramount. DataVault 2.0 is designed to be flexible, and leveraging iterative, agile development techniques allows teams to adapt quickly to changing business requirements. Implementing a continuous integration/continuous deployment (CI/CD) pipeline ensures that changes can be made swiftly and reliably, supporting ongoing data model evolution.

DataVault 2.0 models also benefit from a strong focus on data governance. By adhering to governance principles, teams can ensure data quality and compliance across the architecture. This includes establishing clear data ownership, enforcing security protocols, and tracking data lineage. Data governance ensures that data remains trustworthy and can be easily audited, which is crucial for regulatory compliance and business decision-making.

Effective metadata management is another best practice for DataVault 2.0. Metadata provides valuable insights into how data is structured, where it resides, and how it flows across systems. This helps teams understand the relationships between different data components, making maintaining and troubleshooting the data vault easier. Strong metadata management practices also facilitate reporting, as users can better understand the underlying data structures.

Performance optimization is critical for DataVault 2.0 architectures, especially as data volumes grow. One best practice is to design your system for efficient querying and reporting. For example, denormalizing specific data and creating summary tables in the Gold layer can improve query performance. Another technique is to use indexing and partitioning strategies to enhance the speed of data retrieval, especially when dealing with large-scale datasets.

Another key best practice is to incorporate automation into your DataVault 2.0 pipeline. Automated testing, data quality checks, and data loading processes ensure that the data remains consistent and accurate over time. Automation also reduces the risk of human error and helps accelerate the time to value for data integration efforts. By automating routine tasks, teams can focus on higher-value activities like data analysis and insights generation.

Collaboration between business and technical teams is essential for the success of a DataVault 2.0 implementation. Involving business stakeholders early in the design and development process ensures that the data model aligns with business goals. Regular feedback loops between technical teams and business users can help refine the data vault and ensure it remains aligned with the organization's evolving needs.

Another best practice is monitoring and managing the performance of a DataVault 2.0 implementation. By using monitoring tools and dashboards, teams can track the health of the data vault and identify any issues early. Proactively addressing issues such as slow query performance or data inconsistencies helps maintain the integrity of the data vault and ensures that data is available for decision-making.

Finally, it's essential to maintain a robust documentation framework for your DataVault 2.0 model. Proper documentation makes it easier to onboard new team members, troubleshoot issues, and ensure all stakeholders understand the data model. Clear documentation should include data definitions, lineage, the architecture of hubs, links, and satellites, and any business rules or transformations applied to the data.

Use Case: DataVault 2.0 in the Healthcare Industry

In the healthcare industry, DataVault 2.0 provides an efficient, scalable way to manage vast amounts of healthcare data from various sources, ensuring data consistency, security, and historical accuracy. The architecture's flexibility allows healthcare organizations to integrate data from disparate systems like electronic health records (EHR), insurance claims, and lab results while adhering to stringent regulatory requirements such as HIPAA.

In a typical healthcare DataVault 2.0 implementation, hubs might represent core business entities like patients, doctors, and hospitals. For example, a Patient hub would contain a unique patient identifier (PatientID) as the business key, and no descriptive information would be stored here, adhering to DataVault principles of separation of business keys from descriptive data. This ensures that the patient's identity is captured without risk of exposing personal information.

Links in the healthcare industry typically represent relationships between these hubs. A Patient-Doctor link, for example, would capture the relationship between patients and the doctors who treat them. This link would include keys for both the Patient hub and the Doctor hub and a timestamp to track when the relationship was formed. Links enable the model to manage complex relationships between entities while maintaining a simple structure.

Satellites store descriptive and historical data about the entities represented by the hubs. For example, a Patient satellite would store descriptive information about a patient, such as their name, date of birth, contact information, and other demographic details. Additionally, it could include history-related fields like the patient's address changes or insurance provider details, with each record tagged with a timestamp to track changes over time.

Another example of a satellite in the healthcare industry could be a Doctor satellite, where historical information about the doctor's specialty, qualifications, and work history could be tracked. This allows the organization to retain a complete history of a doctor's career, such as any changes in their specialization or their work location, providing a clear, auditable trail of their professional journey.

A particularly important link in healthcare DataVault 2.0 architectures is the Patient-Visit link, which connects the Patient hub and Visit hub. Each record in this link table can contain a reference to a patient's visit to a healthcare provider, with associated data such as the visit date and the department or service involved. By linking patient visits to the Patient hub, the healthcare provider can create a longitudinal view of patient care.

Within each satellite, tracking the clinical data associated with a visit is also possible. For example, a Visit Clinical satellite could store detailed information such as diagnosis codes (ICD codes), procedures performed (CPT codes), and the medications prescribed during the visit. By structuring these records in a satellite, organizations can keep an accurate historical record of every patient's clinical encounters.

DataVault 2.0's adaptability shines when integrating new data sources into the system. For example, the healthcare provider may add a new data stream such as telemedicine appointments. This new data could be modeled as a new Visit Type hub, with a Telemedicine-Visit link joining it to the Patient and Doctor hubs. This flexibility allows new technologies and data sources to seamlessly integrate into the model without disrupting existing processes.

Another key advantage of DataVault 2.0 in healthcare is its ability to track changes over time. As new regulations emerge or patient care guidelines shift, new data can be added to the architecture without disrupting historical records. For example, suppose new treatment protocols for diabetes are introduced. In that case, the model can easily accommodate these changes by adding new satellites or updating existing ones while keeping all previous data intact.

In addition to managing structured data like patient demographics and clinical records, DataVault 2.0 can also handle unstructured data, such as doctors' notes or medical images. By incorporating technologies such as natural language processing (NLP) and integrating them with the architecture, healthcare organizations can gain deeper insights from text data. For instance, a Doctor's Notes satellite could store raw text extracted from EHR notes, which could later be analyzed for trends or medical insights.

DataVault 2.0's ability to maintain data lineage and version control is invaluable in the healthcare industry. This feature ensures that organizations can track data changes over time, which is critical for regulatory compliance, auditability, and reporting. The Audit satellite can be used to track changes made to sensitive healthcare data, including who made the changes and when, ensuring a fully traceable data history.

By using DataVault 2.0 to model healthcare data, organizations can reduce data silos and create a unified view of a patient's healthcare journey, leading to better outcomes and more personalized care. Integrating diverse data sources—whether clinical, operational, or external—creates a comprehensive data ecosystem that provides the insights needed to improve decision-making and patient care.

With its structured approach and adaptability, DataVault 2.0 is an ideal solution for managing the complex data requirements of modern healthcare organizations. Whether dealing with EHRs, insurance claims, lab results, or new telehealth data, DataVault 2.0's hubs, links, and satellites structure allows healthcare providers to manage their data effectively, enabling more intelligent, more data-driven decisions.

Conclusion

DataVault 2.0 offers a structured, scalable, and flexible framework for managing healthcare data. It effectively addresses the complexity of integrating and processing diverse data sources while ensuring compliance with regulatory requirements. The architecture's use of hubs, links, and satellites provides a straightforward and efficient way to capture and track critical business entities, their relationships, and historical data, enabling healthcare organizations to gain valuable insights.

By leveraging DataVault 2.0, healthcare providers can ensure that patient data is captured accurately, stored securely, and made readily available for decision-making, analytics, and regulatory reporting. The flexibility to adapt to new data sources, such as telemedicine or wearable devices, ensures that the architecture remains future-proof and capable of accommodating emerging technologies.

Ultimately, implementing DataVault 2.0 in healthcare provides a robust, long-term solution for managing data across the organization. Enabling a unified view of the patient journey and supporting data integration foster better outcomes, more personalized care, and improved operational efficiency.

CHAPTER 26

Combining Medallion with DataVault 2.0 and Kimball

This chapter explores how to effectively combine three robust data modeling architectures: Medallion, DataVault 2.0, and Kimball. Each of these frameworks has its unique strengths, and when integrated, they provide a strong and scalable approach to data management. The focus will be on using Medallion Architecture for raw data storage in the Bronze layer, DataVault 2.0 for transforming the Silver layer, and Kimball for business-ready insights in the Gold layer.

We'll explore how the Bronze, Silver, and Gold layers can work together. We will leverage the power of DataVault 2.0 to ensure a scalable and flexible data transformation process while using Kimball to structure the data to facilitate actionable business intelligence. This integration enables seamless data flow, from raw data ingestion to refined business insights.

By combining the strengths of these architectures, organizations can build an end-to-end data pipeline that balances flexibility, scalability, and analytics-driven results. In the following sections, we'll explain how each layer functions in this combined approach, followed by a practical use case in the finance industry.

R. L. Steelman, Jr., *Mastering Snowflake DataOps with DataOps.live*, https://doi.org/10.1007/979-8-8688-1754-0_26

Combining Medallion, DataVault 2.0, and Kimball Architectures

Combining Medallion, DataVault 2.0, and Kimball architectures offers a unified framework that maximizes the strengths of each approach. The Medallion Architecture's three-layer approach works well as the foundational structure for ingesting and transforming data. In this setup, the Bronze layer serves as the raw data lake, offering an unaltered and comprehensive data view. This ensures data governance is in place, with all incoming data recorded without transformations.

The Silver layer introduces DataVault 2.0's powerful transformation and data modeling capabilities. DataVault 2.0 enables organizations to easily manage and link disparate data sources while providing flexibility and scalability. It focuses on maintaining a historical data record, ensuring that changes over time are captured accurately. This layer enhances the Medallion Architecture by offering the necessary structure and data integrity, ensuring that complex data relationships are stored in a way that facilitates easy analysis.

In the Gold layer, the Kimball approach excels by offering a dimensional model that is optimized for analytical querying. Kimball's star schema and fact tables allow for fast aggregation and reporting, providing business users with actionable insights. By applying Kimball here, organizations can create a transparent and efficient data structure for business intelligence, reporting, and analysis.

Each architecture contributes to the data pipeline in different ways, with the Medallion Architecture providing the structure, DataVault 2.0 enabling flexible data transformation, and Kimball delivering a high-performance analytics environment. Together, they create a robust pipeline that allows users to efficiently work with large, diverse datasets while maintaining flexibility and scalability.

This combination also ensures businesses can leverage the advantages of historical tracking and real-time analytics. The Medallion Architecture's raw data storage enables long-term retention and data governance, while DataVault 2.0 ensures that the transformation processes are adaptable to changes in data structure. Meanwhile, Kimball's focus on fast, meaningful analytics ensures that the final data is optimized for decision-making.

With each architecture designed to fulfill a specific purpose, this approach minimizes friction between data ingestion, transformation, and business intelligence. By breaking down data into manageable stages, organizations can adopt a more organized approach that supports complex analytics without sacrificing performance or scalability.

By embracing these combined architectures, businesses can adapt to changing needs and ensure that data is always ready for consumption. With Medallion, DataVault 2.0, and Kimball working together, organizations can build a powerful data pipeline capable of handling vast amounts of data while still providing meaningful and actionable insights at every stage.

DataVault 2.0 in the Silver Layer

The Silver layer in the combined Medallion, DataVault 2.0, and Kimball architecture is critical for transforming raw data into a more usable format, and DataVault 2.0 excels in this role. DataVault 2.0's flexibility in this layer allows organizations to integrate disparate data sources while maintaining an auditable and historical record of all changes. DataVault's focus on creating Hubs, Links, and Satellites helps to organize data and establish connections between various datasets, making it easier to work with complex, structured information.

By using DataVault 2.0 in the Silver layer, organizations can efficiently handle raw, untransformed data, ensuring that it can be easily accessed and integrated into downstream systems. The Hubs, which contain business keys, serve as the foundation for all data relationships, while Links allow for the creation of meaningful associations between entities. This architecture is especially valuable when data from multiple systems or sources needs to be unified, ensuring that users can track and trace how data evolves over time.

DataVault 2.0 allows seamless data transformations that adhere to business rules while preserving data lineage and historical context. The Satellites in this architecture store descriptive information about the data, such as metrics or attributes, capturing changes over time and supporting robust auditing capabilities. This is particularly useful for environments where the data is subject to frequent changes, such as in customer information, product inventory, or sales data.

As part of the Medallion Architecture's Silver layer, DataVault 2.0 introduces a layer of data quality that is essential for businesses relying on clean, accurate datasets. Creating a structured, consistent view of the data ensures that the information available for further transformation or analysis is reliable and fit for use. In this layer, businesses can apply data quality checks to ensure that only the most relevant and accurate data is retained.

The DataVault 2.0 Silver layer also emphasizes flexible design principles, making it easier to accommodate future changes in data sources or models. This adaptability allows for easy integration of new data sources or adjustments to existing systems, giving organizations the agility needed to scale operations and adapt to evolving business needs. By centralizing the data transformation and linking processes in the Silver layer, businesses can create a strong, unified analytics base.

Furthermore, the Silver layer provides the necessary transformations for optimizing the data for business use in the Gold layer. By cleansing, enriching, and linking data through DataVault 2.0, organizations ensure that the information available for analysis is accurate, consistent, and ready to support decision-making. This also facilitates better business intelligence, as the data is more accessible and can be easily aggregated for reporting purposes in the final Gold layer.

DataVault 2.0 in the Silver layer enhances data quality and improves the speed and efficiency of analytics. By ensuring that the data is already linked, cleansed, and ready for use, this layer minimizes the effort required in downstream processing, reducing the time to insight. This is especially critical for organizations that need to respond quickly to changing market conditions or business requirements.

By incorporating DataVault 2.0 into the Silver layer of the combined Medallion Architecture, businesses can create a resilient, scalable data pipeline that supports historical tracking and current data transformation. This powerful combination ensures that data is organized to maximize its value across the entire data lifecycle, from raw ingestion to business intelligence.

Using Kimball in the Gold Layer

In the Gold layer, the Kimball methodology organizes and structures data for business users, providing the ultimate business intelligence layer. This layer focuses on transforming data into a format that is easy to understand, analyze, and use for decision-making. Kimball's focus on dimensional modeling, specifically through the use of fact and dimension tables, ensures that the data is structured for quick and efficient reporting and querying.

Kimball's approach uses star and snowflake schemas, which are highly optimized for query performance and ease of understanding. In the Gold layer, business users often rely on these models to answer key questions, such as sales performance, customer behavior, or operational efficiency. The fact tables store the quantitative measures (such as sales amount or order quantity), while dimension tables store descriptive information (such as customer name or product category).

One key aspect of Kimball modeling in the Gold layer is its focus on simplicity. Kimball's dimensional models are designed to be intuitive, making it easier for non-technical users to interact with and extract insights from the data. This reduces the dependency on IT for querying the data, empowering business users to perform their own analysis, while ensuring that the data is structured in a way that promotes accuracy and efficiency.

Kimball advocates using conformed dimensions to ensure effective use of the Gold layer. These are shared dimensions used across multiple fact tables and subject areas, ensuring consistency in reporting and analysis. For example, a customer dimension may be used in both a sales fact table and a customer service fact table, ensuring that both areas refer to the same customer attributes in the same way.

The use of Kimball in the Gold layer also enhances the performance of analytical queries. By organizing data in a dimensional model, users can query the data quickly and efficiently, with optimized joins between fact and dimension tables. This is particularly beneficial for business intelligence tools and dashboards, which often require fast response times to provide real-time insights to decision-makers.

Furthermore, the Kimball methodology is designed to be flexible and scalable. As the business grows and new data sources are added, the Gold layer can easily be expanded to accommodate new dimensions and facts. This ensures that the architecture can evolve with the business, maintaining a high level of performance and usability even as data volume and complexity increase.

In the Gold layer, the Kimball approach supports the creation of data marts, which are subsets of the data warehouse designed for specific business areas or departments. These data marts allow business users to access a curated set of data that is relevant to their specific needs. By combining the power of Kimball's dimensional modeling with the data governance and transformation capabilities of the Silver and Bronze layers, the Gold layer provides a comprehensive, flexible, and powerful foundation for business analysis and decision-making.

Best Practices for Integrating Datavault 2.0 with the Medallion Architecture

Integrating DataVault 2.0 with the Medallion Architecture presents a powerful approach to managing data in modern environments. By combining these two methodologies, organizations can take advantage of their strengths. DataVault 2.0's flexibility and scalability complement the Medallion Architecture's focus on raw data ingestion, cleaning, enrichment, and analysis. This combination allows for the seamless data flow from raw, unstructured forms into structured, accessible insights for business users.

The integration starts with the Bronze layer, where the raw, untransformed data is ingested. In this layer, DataVault 2.0 shines by using Hubs to uniquely identify business keys, Links to define relationships, and Satellites to track changes over time. The Medallion Architecture provides a solid foundation for ingesting and storing this raw data before it is further processed into the Silver layer.

The Silver layer is where DataVault 2.0 truly adds value. By implementing DataVault 2.0 in the Silver layer, you can organize, cleanse, and enrich the raw data from the Bronze layer. The flexibility of DataVault 2.0 allows for tracking historical data and building relationships across various sources. It also provides an agile approach to handle changes in the data structure, making it ideal for the ongoing transformation needed to populate the Gold layer.

The Silver layer's DataVault 2.0 implementation ensures that data is clean, consistent, and well-governed before it moves into the Gold layer. DataVault 2.0 enables organizations to build a data environment that can evolve over time by focusing on scalability and historical tracking. It allows businesses to manage complex data sets while maintaining a high degree of flexibility in their analytical processes.

In the Gold layer, the Medallion Architecture's focus on analytics and business insights aligns well with the use of Kimball's dimensional modeling. Here, you can integrate DataVault 2.0's cleaned and enriched data to form the basis for building star or snowflake schemas. By adhering to the Kimball approach, businesses can organize their data into fact and dimension tables that are optimized for reporting and analysis.

A key best practice when integrating DataVault 2.0 with the Medallion Architecture is maintaining data lineage. Understanding how data flows from the raw, unstructured form in the Bronze layer through the Silver layer's cleansing and transformation, and finally to the Gold layer for analysis is crucial for ensuring transparency and data governance. Proper data lineage also supports auditing and helps trace issues back to their root cause.

Another best practice is ensuring that the DataVault 2.0 models remain flexible enough to accommodate future business requirements or changes in data sources. Organizations can quickly adapt to changes without disrupting the entire data pipeline by using Hubs, Links, and Satellites. This is especially important in fast-moving industries where new data sources and requirements emerge regularly.

When integrating DataVault 2.0 with Medallion, automation becomes a key factor in achieving efficiency. By automating the process of ingesting raw data in the Bronze layer, transforming it in the Silver layer, and preparing it for analysis in the Gold layer, organizations can significantly reduce the manual effort involved in managing their data. Tools like DataOps.live provide capabilities to automate the workflows that keep data flowing smoothly across all layers.

Lastly, continuous monitoring and optimization are essential for maintaining the performance of a Medallion and DataVault 2.0 integration. As data volumes grow and business requirements evolve, it's vital to regularly assess the architecture for bottlenecks and inefficiencies. Organizations can use monitoring tools to ensure that each layer performs as expected and that data flows smoothly across the entire pipeline. This will help maintain optimal performance and scalability over time.

Optimizing Data Flow from Bronze to Gold: A Unified Approach

Optimizing data flow from the Bronze to the Gold layer in a Medallion and DataVault 2.0 combined architecture is critical for ensuring the overall data pipeline's efficiency, accuracy, and scalability. A unified approach ensures that data moves seamlessly through each stage without bottlenecks, delays, or quality issues, while meeting analytical needs and business requirements.

The first step in optimizing data flow is focusing on the ingestion process in the Bronze layer. Raw data often comes from various sources, such as transactional databases, external APIs, or even real-time streams. Efficient ingestion processes that automate the acquisition and storage of raw data are essential to streamlining subsequent steps. Using tools and platforms like DataOps.live, organizations can automate much of this process, allowing data to flow into the Bronze layer with minimal manual intervention.

Once the data is in the Bronze layer, it is essential to maintain flexibility and scalability. This is where DataVault 2.0's approach to Hubs, Links, and Satellites comes into play. Hubs help manage business keys uniquely, Links represent relationships between those keys, and Satellites store historical information. This modular approach makes scaling easier as new data sources or requirements emerge. DataVault's flexible design is ideal for handling rapidly changing data in environments where new sources of information are continuously added.

Data quality and consistency become paramount in the Silver layer, which implements DataVault 2.0 principles. Transformations like data cleaning, enrichment, and validation should be performed to ensure that only high-quality data moves forward. By leveraging DataVault 2.0's satellite tables, businesses can track historical changes and maintain a complete record of transformations. With tools for automating these processes, companies can minimize manual error and ensure that only relevant, accurate data is made available for further processing.

At the same time, the Silver layer should remain efficient. To do this, organizations should focus on optimizing transformation logic. Rather than transforming all data in real-time, companies can batch process data where applicable or use incremental updates. This approach reduces the load on the system and ensures that data flows smoothly to the Gold layer for analytical processing. Automating transformations, where possible, will also speed up the data preparation process and reduce manual intervention, leading to faster insights.

In the Gold layer, which follows Kimball's dimensional modeling principles, the primary goal is to ensure that the data is structured in a way that is easily consumable by business intelligence tools. This layer typically consists of star or snowflake schemas, which organize facts and dimensions in a manner that enhances reporting and analysis. Optimizing the Gold layer requires focusing on building well-designed facts and dimensions that serve specific business needs while ensuring they are scalable for future reporting requirements.

Integrating data from the Silver layer into the Gold layer can be challenging, especially when working with diverse data sources. One key to overcoming this challenge is ensuring the transformation logic in the Silver layer aligns with the needs of the Gold layer. Additionally, maintaining metadata and data lineage from Bronze to Gold is crucial. This ensures transparency and allows data analysts and data scientists to trace the origin of the data and verify its consistency and reliability.

Another aspect of optimization is maintaining a balance between data storage and access speed. By using a combination of partitioning strategies, indexing, and data caching, the Gold layer can be optimized for quick access by business users and reporting tools. Optimizing data flow through the use of performance-tuning techniques ensures that queries run faster, even as data volumes increase.

Finally, continuous monitoring and testing are essential for keeping the data pipeline running smoothly. Automated monitoring tools can track data flow from one layer to another, checking for issues like bottlenecks or unexpected data quality problems. By setting up automated alerts and performance checks, organizations can detect issues in real time and take immediate action to resolve them before they affect business users. Periodic testing and validation ensure the accuracy and reliability of data as it moves through the pipeline.

Furthermore, using version control and audit trails for the data transformation process enhances governance and traceability. This allows organizations to roll back changes when necessary and maintain a high degree of confidence in the data they deliver. Ensuring that all changes to the data pipeline are fully documented and tracked supports accountability and helps mitigate risks associated with data errors.

As data volume grows, scaling the architecture becomes essential to maintaining optimized performance. In the context of Medallion and DataVault 2.0, this means focusing on both storage and compute scalability. Cloud-based solutions, which offer elastic scaling, make it easier to handle growing data sets without compromising performance. Ensuring that the architecture can scale efficiently is crucial for future-proofing the system as the volume of data continues to increase.

Finally, companies should integrate proactive error-handling processes to make the entire pipeline more resilient and efficient. Automated exception handling in both the Silver and Gold layers can ensure that issues like missing data or unexpected format changes are caught early. This minimizes downtime and ensures that the data flow remains smooth and uninterrupted. With the right error-handling strategy in place, businesses can maintain the integrity of their data pipeline and provide reliable reporting and analysis.

Managing and Scaling the Combined Architecture

Managing and scaling the combined Medallion, DataVault 2.0, and Kimball architecture requires a structured approach to ensure that the data pipeline remains efficient, scalable, and adaptable. It begins with clearly defining the roles of each layer and understanding how data flows through them. As data volume and complexity increase, so too must the systems and processes that manage them. Organizations must continually assess their existing infrastructure to ensure it can handle growing data demands without compromising performance.

One of the most critical factors in managing this architecture is ensuring that each layer—Bronze, Silver, and Gold—can scale independently. In practice, this means ensuring that raw data ingestion in the Bronze layer can handle large data volumes without overwhelming storage or compute resources. For the Silver layer, which builds upon the raw data using DataVault 2.0, scaling involves ensuring that transformations, such as linking and satellite updates, can be performed at scale without creating bottlenecks. The Gold layer must also be scalable to support efficient reporting and analysis.

Automation plays a central role in scaling the combined architecture. Automating data ingestion, transformations, and delivery is crucial with the integration of Medallion, DataVault 2.0, and Kimball. Leveraging automation tools in each layer helps to streamline processes, reduce errors, and speed up the delivery of valuable insights to business users. For example, automating the transformation of raw data in the Bronze layer to DataVault 2.0's Silver layer ensures that data processing is efficient and up-to-date. In contrast, automated data aggregation in the Gold layer helps to create business-ready reports quickly.

Organizations must also focus on performance optimization as the data architecture scales to ensure efficient query processing in the Gold layer. Implementing best practices like indexing, partitioning, and caching will improve the performance of data retrieval and reporting. Furthermore, as data grows, ensuring that data models in the Silver and Gold layers remain optimized for query performance is key to avoiding long query times and inefficient processes. This requires ongoing system performance monitoring and fine-tuning of the database and data models.

Data governance and security also become increasingly important as the architecture scales. Organizations must implement robust policies for managing data access and ensuring that sensitive data is protected. With the combined architecture handling large volumes of sensitive and valuable business data, implementing access control mechanisms and monitoring tools will help prevent unauthorized access and ensure compliance with regulations. A well-structured governance framework ensures that data integrity is maintained across all layers.

Finally, it is important to adopt a flexible and adaptable approach to scaling the architecture over time. The combined Medallion, DataVault 2.0, and Kimball architecture should evolve as business needs and data sources change. This might involve adding new data pipelines, implementing more granular data models, or expanding the use of automation tools to meet the demands of growing data volumes. Organizations can effectively manage and scale this sophisticated data architecture by maintaining flexibility and staying proactive in evaluating performance and governance requirements.

Use Case: Combining Medallion, DataVault 2.0, and Kimball in the Finance Industry

The combination of Medallion, DataVault 2.0, and Kimball architectures in the finance industry creates a robust framework for managing vast amounts of transactional and analytical data. In this use case, we can explore how each layer contributes to creating a robust and scalable data pipeline that supports the needs of financial institutions. By organizing data into the Bronze, Silver, and Gold layers, the architecture ensures high data quality, detailed business insights, and agile reporting that are critical in the fast-paced world of finance.

The Bronze layer serves as the foundation of the data pipeline. For a financial institution, this layer stores raw transactional data such as trades, account movements, and financial market data. This unrefined, raw data might be sourced from multiple channels like banking systems, payment gateways, or market feeds. Medallion Architecture's focus on creating a raw data lake in the Bronze layer allows the financial institution to capture this information without any transformations, ensuring that data remains available for further enrichment in later stages. A strong focus on data quality is vital, as financial transactions must remain traceable and auditable at all stages.

Once the data has been captured, it is processed into the Silver layer, leveraging DataVault 2.0's strengths for detailed structuring and modeling. In this layer, raw data is enriched and transformed into valuable business entities that align with the financial institution's operations. For example, customer details, transactional events, and accounts can be stored in the form of Hubs, Links, and Satellites. A hub table might contain a unique identifier for each customer, while a link table stores the relationships between customers and their financial transactions. Satellite tables, in turn, contain historical attributes, such as transaction amounts, dates, and financial instruments. By organizing data in this way, financial institutions can track the history and relationships of every transaction over time, providing a complete audit trail.

The Gold layer in the combined architecture is where business-ready insights are created for reporting and decision-making. Using the Kimball approach, financial institutions can aggregate and transform the structured data into dimensional models suitable for business analysis. This layer can calculate and present key performance indicators (KPIs) such as monthly revenue, portfolio performance, or loan default rates in easy-to-understand formats. For example, a financial institution might build a data mart for reporting on customer profitability, where facts like total transactions, loan balances, and associated dimensions such as customer demographics, region, and product type help analysts quickly assess trends and make data-driven decisions.

Implementing Medallion with DataVault 2.0 and Kimball in the finance industry also optimizes data flow and accessibility. Financial analysts and data scientists can directly query the Gold layer's optimized dimensional models for performance reports and dashboards. With DataVault 2.0 in the Silver layer handling historical data and transactions, and Kimball's dimensional models in the Gold layer organizing them for business analysis, the data pipeline remains clean, scalable, and agile. Additionally, each layer's focus ensures that data is accessible with minimal delay, even as the volume of financial transactions increases over time.

One key benefit of this combined architecture is the enhanced ability to maintain high data integrity and regulatory compliance. Maintaining a traceable, auditable history of transactions and customer data in the finance industry is critical for meeting regulatory standards. The DataVault 2.0 approach, with its Hubs, Links, and Satellites, ensures that all business transactions are recorded consistently and traceably, preserving the integrity of the financial data throughout its lifecycle. Meanwhile, the Kimball model's focus on reporting allows for the creation of detailed reports that meet compliance requirements without sacrificing performance or scalability.

Moreover, combining these architectures ensures better performance in the long run. Financial institutions often struggle with growing data volumes, especially as they integrate new data sources or scale their operations. With the Medallion, DataVault 2.0, and Kimball framework, each layer is optimized for specific tasks, ensuring that financial data is processed efficiently. The Bronze layer's raw data storage ensures that the system can handle large data volumes, while the Silver and Gold layers efficiently transform and report on that data, minimizing the strain on resources while optimizing performance.

In addition, this architecture supports flexibility and adaptability in the face of changing business requirements. As new regulatory requirements emerge or new financial products are launched, the Medallion Architecture makes it easy to ingest raw data quickly into the Bronze layer. The Silver layer can be easily modified to accommodate new data sources or relationships, while the Gold layer's dimensional models can be updated to reflect new KPIs or business metrics. This adaptability makes the combined Medallion, DataVault 2.0, and Kimball architecture an ideal solution for the dynamic nature of the finance industry.

Another example of the combined architecture's utility is in fraud detection and risk management. With historical data and real-time transaction data stored in the Silver layer's DataVault 2.0 model, financial institutions can easily track activity patterns across various systems. Links between customers, transactions, and accounts enable advanced analytics that can detect unusual patterns and flag potential fraudulent behavior. The Gold layer can aggregate these patterns and display risk metrics to decision-makers, allowing them to take swift action to mitigate risks.

The ability to combine these three powerful architectures also allows for continuous data quality monitoring. In the financial sector, ensuring data integrity is paramount. Financial institutions can monitor data accuracy over time by leveraging DataVault 2.0's consistency and historical tracking. If any discrepancies are found, they can be traced back to the source, ensuring that any data anomalies are promptly addressed, maintaining the accuracy of financial reports, and preventing costly errors or compliance failures.

Furthermore, scalability is an important consideration in the finance industry, where large amounts of transactional data are processed daily. The Medallion, DataVault 2.0, and Kimball architectures scale seamlessly as new data sources are added or as the volume of data grows. The Bronze layer is capable of handling this influx of raw data, while the Silver and Gold layers can adapt as new requirements emerge, all while ensuring that data is processed and stored with speed and efficiency.

In conclusion, the Medallion, DataVault 2.0, and Kimball architecture combination provides a powerful, scalable, and adaptable data pipeline for the finance industry. It balances the need for raw data ingestion, detailed historical tracking, and high-performance reporting while maintaining data integrity and compliance. By implementing this unified architecture, financial institutions can optimize their data flow, create actionable business insights, and make data-driven decisions to stay ahead in a competitive and regulated market.

Conclusion

Bringing together Medallion Architecture, DataVault 2.0, and Kimball modeling creates a structured yet flexible approach to data management, optimizing raw data ingestion, historical tracking, and business intelligence. Organizations can balance scalability with governance by segmenting data into Bronze, Silver, and Gold layers while ensuring that data remains accessible and business-ready. This framework not only supports real-time analytics but also provides the resilience needed to accommodate evolving business and regulatory requirements.

As organizations continue to handle increasing volumes of complex data, adopting a unified architecture becomes even more critical. The synergy between DataVault 2.0's structured historical storage, Kimball's dimensional modeling for analytics, and Medallion's efficient data processing allows businesses to make informed, data-driven decisions. Whether managing financial transactions, customer insights, or risk assessments, this approach ensures that data is both accurate and actionable, streamlining operations while reducing redundancies.

This chapter has demonstrated how combining these methodologies provides a powerful foundation for modern data architectures. By leveraging the strengths of each approach, businesses can create a data ecosystem that is robust, scalable, and adaptable to future demands. Whether in finance, healthcare, or other industries, this model serves as a blueprint for organizations seeking to transform their data strategies while maintaining agility in an increasingly data-driven world.

CHAPTER 27

Entity-Relationship (ER) Modeling and Beyond

Entity-Relationship (ER) modeling has long been a foundational approach to structuring and organizing data, clearly representing entities, their attributes, and the relationships between them. This methodology has been essential in designing relational databases, ensuring data integrity, and facilitating efficient querying. However, as data ecosystems have grown in complexity, traditional ER models have had to evolve to accommodate new requirements such as hierarchical relationships, temporal data, and integrations with modern architectures like data lakes and data warehouses.

This chapter explores both the fundamental principles of ER modeling and its extensions beyond traditional database design. We will examine advanced concepts such as cardinality, normalization, and integrity constraints while also addressing how ER modeling intersects with modern data architectures. Additionally, we will explore how ER models can serve as a bridge between structured relational systems and contemporary frameworks like Data Vault, Medallion, and NoSQL databases. By understanding these concepts, organizations can design data models that are robust and adaptable to the demands of modern analytics and business intelligence.

Fundamentals of ER Modeling: Entities, Attributes, and Relationships

Entity-Relationship (ER) modeling is built upon three core components: entities, attributes, and relationships. Entities represent real-world objects or concepts that are distinguishable within a system, such as customers, products, or transactions. Each entity has attributes, which define its characteristics, such as a customer's

R. L. Steelman, Jr., *Mastering Snowflake DataOps with DataOps.live*, https://doi.org/10.1007/979-8-8688-1754-0_27

name, address, and email. Relationships establish how entities are connected to each other, such as a customer placing an order or a student enrolling in a course. These foundational elements form the basis of a well-structured ER model.

Attributes in ER modeling can be classified into different types, such as simple attributes (atomic values like a name or age), composite attributes (a structured collection like a full name with first and last names), and derived attributes (values that can be computed from other attributes, such as age from a birthdate). Attributes can also be mandatory or optional, helping to define the integrity rules of a database. Defining attributes ensures data is stored efficiently while maintaining accuracy and consistency.

Relationships in ER models establish how entities interact with one another. They are classified based on their cardinality, which describes the number of instances involved in a relationship. The three primary cardinalities are one-to-one (e.g., a passport assigned to a single individual), one-to-many (e.g., a customer placing multiple orders), and many-to-many (e.g., students enrolling in multiple courses). Understanding these relationships is crucial for designing an efficient and normalized database structure.

ER diagrams visually represent entities, attributes, and relationships using standardized notations. Entities are depicted as rectangles, attributes as ovals, and relationships as diamonds, with connecting lines showing how entities are linked. Additionally, primary keys, which uniquely identify each entity, and foreign keys, which establish relationships between entities, play a critical role in maintaining referential integrity. These diagrams provide a blueprint for database development, helping to ensure that data is logically structured.

Normalization is another essential concept in ER modeling, as it helps minimize data redundancy and improves database efficiency. The process involves organizing attributes into separate tables to reduce data duplication and ensure data integrity. Normalization follows a series of rules, called normal forms, that guide how data should be structured. While normalization enhances consistency, it must be balanced with performance considerations, particularly when dealing with complex relationships.

Beyond traditional relational databases, ER modeling principles are also applicable in modern data architectures. While relational systems use structured schemas, NoSQL databases and data lakes often require more flexible data models. Understanding ER modeling helps data architects adapt relational concepts to semi-structured and unstructured data environments, ensuring data is well-structured regardless of the storage technology used.

By mastering the fundamentals of ER modeling—entities, attributes, and relationships—data professionals can create robust database structures that support efficient querying, reporting, and analytics. Whether applied in traditional databases or modern architectures, ER modeling remains a crucial tool for organizing and managing data effectively.

Advanced ER Modeling, Cardinality, Normalization, and Integrity Constraints

Advanced ER modeling builds upon the foundational principles of entities, attributes, and relationships by incorporating more complex structures such as generalization, specialization, and aggregation. Generalization is a process where multiple entity types with shared attributes are combined into a higher-level entity, such as merging "Car" and "Truck" into a generalized "Vehicle" entity. Specialization is the reverse, where an entity is divided into more specific sub-entities, such as breaking "Employee" into "Manager" and "Technician." Aggregation allows an entire relationship to be treated as a higher-level entity, enabling more sophisticated modeling of real-world scenarios.

Cardinality defines the number of occurrences of one entity that can be associated with occurrences of another entity. In advanced ER modeling, concepts like multi-valued dependencies and ternary relationships come into play beyond the standard one-to-one, one-to-many, and many-to-many relationships. A ternary relationship involves three entities simultaneously, such as a "Doctor," "Patient," and "Hospital," representing which doctors treat which patients at specific hospitals. Understanding these complex relationships helps prevent data inconsistencies and ensures accurate data representation.

Normalization is a systematic approach to organizing database structures to reduce redundancy and improve data integrity. It progresses through normal forms, each addressing specific anomalies. First normal form (1NF) eliminates duplicate columns and ensures atomicity, meaning each field contains only one value. Second normal form (2NF) removes partial dependencies, ensuring that attributes depend entirely on a primary key. Third normal form (3NF) eliminates transitive dependencies, where non-key attributes depend on other non-key attributes rather than the primary key. Advanced normalization techniques, such as Boyce-Codd Normal Form (BCNF), ensure even stricter data integrity.

Despite its benefits, normalization must be balanced against performance considerations. Highly normalized databases can lead to excessive joins when querying data, which may impact performance in large-scale applications. In some cases, denormalization—introducing some controlled redundancy—can be strategically used to improve query performance. This is particularly relevant in data warehousing, where retrieving data efficiently for analytics is prioritized over strict normalization rules.

Integrity constraints are vital in enforcing data consistency and accuracy within an ER model. These include entity integrity, which ensures that each table has a unique primary key; referential integrity, which maintains valid relationships between tables through foreign keys; and domain integrity, which ensures attribute values fall within predefined constraints, such as valid date ranges or numerical limits. These constraints help maintain database reliability and prevent anomalies such as orphaned records or invalid data entries.

Incorporating advanced integrity constraints can also include business rules, which define custom conditions that reflect real-world logic. For example, a business rule might state that "a customer must have at least one valid order before being assigned a loyalty tier." Such constraints ensure that the database structure aligns with business operations and enforces logical consistency beyond standard integrity rules.

Advanced ER modeling techniques, combined with proper cardinality definitions, normalization strategies, and integrity constraints, enable the creation of well-structured, high-performance databases. Whether applied in operational databases, data warehouses, or hybrid environments, these principles help ensure that data remains accurate, scalable, and optimized for both transactional and analytical use cases.

Extending ER Models: Incorporating Hierarchies, Temporal Data, and Beyond

Extending ER models beyond traditional entity-relationship structures allows for more sophisticated data modeling, including the incorporation of hierarchies. Hierarchical relationships define parent-child structures where an entity can be associated with multiple subordinate entities. For example, an "Organization" entity might have multiple levels such as "Department," "Team," and "Employee," each forming a hierarchy. These relationships are particularly useful in representing structures such as product categories, company management levels, and geographical regions. Implementing recursive relationships, where an entity references itself, is a common approach for modeling hierarchies within ER diagrams.

Temporal data adds another dimension to ER modeling, allowing databases to track historical changes over time. Traditional ER models capture only the current state of data, but many business applications require the ability to store past values, track trends, and manage effective dates. Temporal modeling techniques introduce valid-time and transaction-time attributes to entities, enabling the creation of time-aware data structures. For instance, an "Employee" entity could include fields for "Start Date" and "End Date" to reflect job history, allowing queries to retrieve historical employment records.

Incorporating temporal data requires careful design to ensure efficiency and integrity. One approach is to use slowly changing dimensions (SCDs), commonly applied in data warehousing to manage attribute changes over time. Type 1 SCDs overwrite old data, while Type 2 SCDs maintain history by creating a new record with a timestamp. Type 3 SCDs store limited history by adding an "Old Value" attribute within the same row. Choosing the appropriate method depends on business requirements and query performance considerations.

Beyond hierarchies and temporal extensions, modern ER models also integrate complex data structures such as semi-structured and unstructured data. Traditional relational databases primarily handle structured data, but with the rise of big data and NoSQL databases, ER modeling has adapted to incorporate flexible storage solutions. Entities may now include JSON, XML, or key-value pairs, allowing for dynamic schema evolution. For example, an "E-Commerce Order" entity might store product details as a JSON object, enabling the system to accommodate varied product attributes without schema changes.

Another advanced extension of ER modeling involves polymorphic relationships, where a single foreign key can reference multiple entity types. This is useful in scenarios where different entities share a common relationship. For example, a "Comment" entity in a content management system could be associated with multiple parent entities such as "Article," "Video," or "User Profile." Implementing polymorphism requires strategies like shared primary keys, subtype tables, or foreign key constraints with metadata tracking.

Database architectures can better reflect real-world complexities by extending ER models to include hierarchies, temporal tracking, semi-structured data, and polymorphic relationships. These enhancements allow businesses to build flexible, scalable, and future-proof data models supporting transactional and analytical workloads. As data requirements evolve, modern ER modeling approaches ensure that relational databases remain relevant in increasingly diverse and dynamic environments.

From ER Models to Modern Data Architectures: Bridging the Gap

Entity-Relationship (ER) models have long been the foundation of structured data design, offering a clear and systematic approach to defining entities, attributes, and relationships. However, as data architectures evolve to accommodate large-scale, distributed, and real-time processing needs, traditional ER modeling must be adapted to fit modern paradigms. Bridging the gap between ER models and contemporary architectures such as data lakes, DataVault 2.0, and cloud-native platforms requires rethinking how data is structured, integrated, and utilized.

One key challenge in transitioning from ER modeling to modern architectures is accommodating the shift from rigid relational schemas to flexible, schema-on-read approaches. Traditional ER models enforce strict normalization, hindering performance in analytical environments. Modern architectures often favor denormalized structures, such as star or snowflake schemas, or hybrid approaches like DataVault, which balance normalization for integration with denormalization for performance optimization. This shift requires re-evaluating how entities and relationships are defined in evolving ecosystems.

Another significant consideration is the role of data integration across diverse sources. ER models excel at representing structured, transactional data. Still, modern architectures must ingest and process semi-structured and unstructured data from sources such as IoT devices, social media, and external APIs. Technologies like Snowflake and Databricks enable organizations to store JSON, XML, and Parquet files alongside structured tables, requiring an extension of ER principles to incorporate flexible data formats while maintaining referential integrity and business context.

Modern data architectures also demand new data governance, security, and lineage tracking approaches. Traditional ER models assume centralized control, but cloud-based and decentralized architectures introduce challenges in maintaining consistency across distributed data environments. Implementing metadata-driven automation, data catalogs, and lineage tracking tools ensures that data remains traceable and trustworthy even as it moves between structured ER models and more dynamic processing layers, such as DataVault hubs and links or event-driven data pipelines.

The need for real-time analytics and streaming data processing further challenges the static nature of ER models. Modern architectures leverage event-driven systems, where data flows continuously rather than being stored in rigid transactional databases. This shift requires integrating ER modeling concepts with event sourcing, stream processing frameworks, and change data capture (CDC) techniques to ensure that relational entities remain relevant while supporting near real-time business intelligence and operational decision-making.

Ultimately, bridging the gap between ER models and modern data architectures requires a hybrid mindset—leveraging the strengths of structured relational modeling while embracing the flexibility and scalability of cloud-native, big data, and analytical frameworks. By integrating ER modeling principles with data lakes, DataVault, and Kimball methodologies, organizations can create robust architectures that support operational efficiency and strategic insights, ensuring long-term adaptability in an ever-evolving data landscape.

Conclusion

Entity-Relationship (ER) modeling remains a fundamental approach to structuring data, but its role in modern data architecture has evolved significantly. While ER models continue to provide clarity in defining relationships between entities, they must now coexist with flexible, scalable frameworks that support real-time analytics, semi-structured data, and cloud-native storage. Organizations that successfully integrate ER principles with modern methodologies, such as Medallion Architecture, DataVault 2.0, and Kimball modeling, can achieve both structured consistency and adaptability to evolving business needs.

As businesses handle increasing volumes of data from diverse sources, the ability to transition from traditional ER models to architectures that support streaming, machine learning, and predictive analytics becomes essential. This shift requires balancing normalization and denormalization, implementing robust data governance, and leveraging automation to maintain data integrity across hybrid environments. By embracing hybrid approaches, organizations can ensure that their data models remain relevant and valuable for operational and analytical use cases.

The future of data modeling lies in the ability to merge foundational principles with innovative frameworks that support agility, scalability, and interoperability. ER modeling provides a strong backbone for relational consistency, but it must be adapted to fit the needs of cloud-based data platforms and real-time processing pipelines. By understanding the strengths and limitations of each approach, businesses can design architectures that not only preserve data quality but also empower advanced analytics and decision-making in an increasingly complex data landscape.

CHAPTER 28

Event-Driven Data Models

Event-driven data models focus on structuring data around events, capturing the occurrence of significant actions or changes within a system. These events drive data flow, ensuring systems can respond in real-time to changes. By emphasizing events over traditional transaction-based models, event-driven architectures provide flexibility and scalability for modern data environments, particularly in industries that require rapid response times and adaptability.

One key advantage of event-driven data models is their ability to support real-time data processing and analytics. Organizing data based on events rather than fixed schedules or transactional updates allows these models to quickly react to new information, providing businesses with up-to-date insights. This is especially valuable in sectors such as e-commerce, finance, and IoT, where timely data access is critical for decision-making.

In the following sections, we will explore how event-driven data models are structured, how they interact with other modern architectures like microservices, and best practices for implementing them effectively. We'll also discuss the role of streaming data platforms, such as Apache Kafka, and how they help create efficient event-driven architecture.

Understanding Event-Driven Architecture: Principles and Core Concepts

Event-driven architecture (EDA) is a design paradigm that focuses on capturing events as the fundamental units of change within a system. An event typically refers to a significant change or action, such as a user interaction, a transaction, or a system state change. In this model, system components react to these events in real-time, triggering responses that can be asynchronous and decoupled from the originating process. This design enables systems to be highly responsive and scalable.

R. L. Steelman, Jr., *Mastering Snowflake DataOps with DataOps.live*, https://doi.org/10.1007/979-8-8688-1754-0_28

One of the core principles of EDA is decoupling. Traditional systems often rely on tightly coupled processes, where one component must wait for another to complete before moving forward. In contrast, EDA promotes a loosely coupled approach where producers of events don't need to know about the consumers. This separation improves flexibility and allows for more dynamic interactions within a system. It also enhances the ability to handle large volumes of data and enables better fault tolerance.

Events in EDA can be thought of as state transitions that represent a change in the system. For example, when a customer places an order, an event stating that the "OrderPlaced" event has occurred might be generated. This event can be processed by multiple systems that handle different tasks: inventory, shipping, and payment. This allows for parallel processing and faster reaction times. Events serve as the foundational building blocks that trigger processes across various domains, making them a critical aspect of modern data architectures.

EDA heavily relies on messaging systems to transfer events between components. These systems can be designed to ensure reliable delivery, manage retries, and handle out-of-order events. Popular messaging platforms such as Apache Kafka or AWS EventBridge provide scalable solutions for managing and distributing events. They act as the backbone of EDA by ensuring that events can be transmitted in real time and handled appropriately by the system.

Another key concept of EDA is event sourcing. In this approach, the system stores each event in a sequence, rather than just the system's current state. This means that by replaying the events, you can reconstruct the system's state at any given point in time. This provides an audit trail and helps in troubleshooting and data recovery. Event sourcing offers better traceability, making it especially valuable for systems that require transparency and accountability.

Event-driven models also leverage the concept of event streams. An event stream is a continuous flow of events that are processed in real-time. Event streams allow businesses to act on data as it is produced, providing them with up-to-the-minute insights and the ability to make data-driven decisions instantly. Streaming platforms, like Apache Flink or Spark Streaming, can be integrated with event-driven architectures to handle large-scale, high-throughput event processing.

To successfully implement an event-driven architecture, it is crucial to ensure that events are well-defined and that proper event schemas are established. Standardized event formats help ensure that different system components can communicate effectively. Furthermore, event versioning becomes essential as the system evolves, providing backward compatibility while allowing for new features and changes to be introduced without breaking existing processes.

Overall, event-driven architecture is increasingly becoming the preferred choice for businesses needing real-time data and requiring flexible, scalable systems. It allows organizations to be agile, responsive, and better equipped to deal with the complexities of modern data environments. With its focus on events, EDA is a powerful paradigm that supports rapid innovation, higher data throughput, and more efficient data processing.

Designing Event-Driven Data Models: Events, Streams, and State Changes

Designing event-driven data models begins with understanding that events, streams, and state changes are at the heart of the architecture. An event represents a significant change in the system, and the data model must reflect this dynamic nature. Unlike traditional data models, which are built around static tables and entities, event-driven models must be designed to handle continuous, real-time changes. Events represent these changes and are processed in sequences, allowing for a more flexible and scalable architecture. The system needs to be built around capturing, storing, and processing events as they occur.

The difference between events and streams is an important concept in event-driven data models. While an event is a single, discrete occurrence that signals a change in the system's state, a stream is a continuous flow of events over time. Streams are used to represent sequences of events that are related in context. Designing with streams requires ensuring the system can handle vast amounts of data flowing through in real-time while maintaining data integrity and order. Each event in the stream adds new data that might influence the system's state, and these must be captured and processed effectively.

The design of event-driven data models also heavily involves handling state changes. In traditional databases, state is typically stored in rows and columns, reflecting the current state of entities. However, in an event-driven model, state is often the result of a sequence of events. For instance, an e-commerce platform might store an event for an item being added to a cart, another event for the customer proceeding to checkout, and a third event for the completion of the purchase. The final state—the fact that the purchase was made—can be determined by aggregating and processing these individual events.

Event-driven models must incorporate mechanisms for maintaining consistency and reliability to ensure these state changes are managed effectively. In many cases, event sourcing stores events sequentially so that the state can be rebuilt at any point by replaying the events in the order they were originally recorded. This ensures that all historical changes are accounted for and that the state is not reliant on a snapshot but rather on an immutable log of events. This approach also provides traceability and an audit trail for system activities.

One of the challenges in designing event-driven models is managing the schema of events. Each event typically has a schema that defines its structure, including the fields, types, and data relationships. Designing effective event schemas ensures compatibility between event producers and consumers. Schemas should be designed to be extensible, as event-driven systems often evolve over time. Careful attention to versioning and schema management can prevent breaking changes from impacting the flow of events and ensure that both old and new versions of events can coexist.

Event-driven data models also involve managing the flow of events across different systems and services. Events may be generated by various sources—such as applications, sensors, or users—and consumed by different components of the system. For instance, an event that signals a new order placed by a customer may trigger actions such as inventory management, payment processing, and order shipping. Each of these actions may require its own processing logic, but the event itself serves as the central piece of data that binds them all together.

Designing effective event-driven data models also requires considering the impact of time. Some events are time-sensitive, meaning they need to be processed in real-time, while others may be processed in batches. For example, in a financial system, a market price event must be processed immediately to reflect changes in stock prices. In contrast, transaction logs might be processed in batches at regular intervals. Incorporating time-sensitive processing into the model requires careful attention to ensure that events are processed according to their urgency.

Moreover, as event-driven systems grow in complexity, defining the relationships between events and services becomes important. These relationships often manifest as event workflows, where one event triggers a sequence of actions across various services. For example, an event may trigger the start of a complex process that involves multiple events and actions, forming a chain of dependencies. Understanding how events interact and depend on each other helps ensure the data flows correctly and efficiently through the system.

Finally, monitoring and tracking events across a distributed system are essential to maintaining performance, reliability, and data integrity. Tools like event stream processors and observability platforms can monitor the flow of events, detect anomalies, and ensure that all events are processed successfully. This is critical in a real-time system where delays or failures in event processing can lead to incorrect state changes, ultimately affecting business operations.

Integrating Event-Driven Models with Data Warehouses and Data Lakes

Integrating event-driven models with data warehouses and lakes involves bridging real-time data processing with traditional batch processing systems. While data lakes and warehouses typically handle batch updates and large-scale data storage, event-driven models process continuous data streams, making them ideal for real-time applications. The challenge lies in ensuring that event-driven data flows seamlessly into the structured environments of data warehouses and lakes without losing the dynamic nature of the data.

One approach to integration is to use event streaming platforms like Kafka or AWS Kinesis, which serve as intermediaries between event-driven systems and traditional data storage. These platforms allow events to be ingested in real-time, processed, and then stored in a data lake for further analysis. Data lakes, known for their flexibility in handling unstructured data, are well-suited to store raw events and allow for downstream processing. The integration ensures that data, regardless of format or source, can be captured and retained efficiently.

Data warehouses, on the other hand, often require data to be transformed into a more structured format. For effective integration, event-driven models can leverage ETL or ELT pipelines to transform the event data into a format compatible with a warehouse's schema. The raw data stored in a data lake can be transformed into aggregated, cleaned, and structured data that can be loaded into the warehouse. These processes may involve using tools like Apache Spark or Snowflake's data processing capabilities to handle structured and semi-structured data.

When integrating with a data warehouse, the timing of the event processing is crucial. Since event-driven systems focus on real-time processing, aligning the timing of batch uploads to the data warehouse is key. Typically, data from events might be processed in near real-time or on a micro-batch schedule, ensuring that updates reflect the most recent events without overwhelming the data warehouse with excessive requests. Efficient synchronization between the systems reduces latency and enhances overall performance.

In addition, handling the schema evolution of event-driven systems when integrating with data lakes or warehouses is a challenge. Events may evolve over time, requiring systems to adapt to changing structures or new event types. This is where tools like schema registries come into play. These registries can help maintain backward compatibility and allow the seamless addition of new event attributes while preserving the integrity of existing data structures in both data lakes and warehouses.

Lastly, integrating event-driven data models with these storage systems enables more advanced analytics. With a unified approach, data analysts and data scientists can run queries across historical data stored in data lakes and up-to-date event data. This combination allows for more dynamic and real-time insights, enabling organizations to make faster, more informed decisions.

Best Practices for Implementing Event-Driven Data in Modern Architectures

Implementing event-driven data in modern architectures requires careful planning to ensure that real-time data flow is consistent and scalable. First, it is crucial to design an architecture that can handle high volumes of events with minimal latency. Event streams should be processed and ingested quickly, with systems designed to scale horizontally as the volume of events grows. This ensures that the system remains performant as demand increases and as more events are generated across various sources.

A critical best practice is to use event streaming platforms like Kafka, RabbitMQ, or AWS Kinesis to handle event streams. These platforms facilitate decoupling the producer and consumer of events, ensuring the system remains flexible and fault-tolerant. Message brokers allow you to manage and monitor event flows more efficiently, allowing for retries and event durability without risking data loss. These platforms also allow for real-time processing, which is essential for keeping the data flow consistent with minimal delay.

Implementing idempotency in event processing is essential to ensure that the system remains robust. Since events can be reprocessed due to failures or retries, ensuring the event processing is idempotent prevents duplicate data from being ingested. Idempotency guarantees that repeated events don't cause inconsistencies in downstream data systems, which is particularly important when integrating with data lakes and warehouses. Properly designing the event consumers to be idempotent also simplifies the architecture and reduces the complexity of handling errors.

Event schema management is another best practice when implementing event-driven data models. Since event data often evolves over time, it's essential to establish a versioned schema management system. Tools like Confluent Schema Registry can help you manage different versions of your event schema, ensuring backward compatibility and preventing the breaking of downstream consumers. This practice is particularly critical in large, distributed systems where events come from multiple sources, and schema changes must be handled gracefully to avoid data inconsistencies.

It's also crucial to have clear data governance policies in place. As real-time event-driven systems often involve massive amounts of data, managing data quality, security, and compliance becomes even more critical. Implementing proper data lineage, tracking where and how events are processed, and ensuring that sensitive data is properly handled are all best practices that help maintain control and integrity within the system. A well-defined governance framework ensures that the architecture complies with legal and business requirements and is transparent across all stakeholders.

Another best practice is to design your event-driven system with fault tolerance in mind. This includes ensuring that events are reliably stored and can be reprocessed in case of failure. One way to achieve this is by leveraging event sourcing and event replay, where events are stored as an immutable log and can be replayed to reconstruct the system's state. This approach ensures that data loss is minimized and the system can recover quickly from failures without losing valuable event data.

Finally, monitoring and observability should be integral to your event-driven architecture. Setting up comprehensive logging, monitoring, and alerting for event processing allows teams to track the system's health in real-time. Metrics such as event processing time, error rates, and throughput can help identify bottlenecks or failures early. With the dynamic nature of event-driven systems, proactive monitoring ensures that any issues can be quickly addressed before they escalate, maintaining the reliability of your system over time.

Conclusion

Event-driven data architectures revolutionize how organizations process and respond to real-time data. With the increasing complexity of business needs and the growing volume of data, these systems provide a flexible and scalable approach to capturing, processing, and analyzing data from various sources. By focusing on events rather than static data, businesses can respond more dynamically to changes and maintain continuous operational flow.

One key benefit of event-driven architectures is the ability to decouple data producers from consumers, allowing for greater flexibility and scalability. This enables systems to process real-time events, reducing latency and improving responsiveness. Whether you're handling financial transactions, sensor data, or customer interactions, the real-time nature of event-driven models can provide immediate insights and drive more informed decision-making.

However, building robust event-driven systems comes with challenges, particularly when integrating with traditional data warehouses and lakes. Ensuring data consistency, managing schema evolution, and implementing real-time data governance are some of the crucial aspects that must be carefully considered. By following best practices and adopting the right tools, organizations can mitigate these challenges and unlock the full potential of event-driven architectures.

As businesses continue to rely on data-driven decision-making, event-driven models will play an increasingly central role in modern architectures. By embracing these principles, organizations can ensure that their data systems are agile, efficient, and ready to meet future demands. The evolution of these architectures will undoubtedly shape the future of data management, empowering businesses to gain real-time insights and stay competitive in an ever-changing marketplace.

CHAPTER 29

Graph Data Modeling

Graph data modeling is an innovative approach for structuring and querying data that excels in scenarios where relationships between entities are complex and dynamic. Unlike traditional relational databases, which use tables to store data, graph databases represent data using nodes, edges, and properties. These structures are ideal for modeling networks, social media relationships, fraud detection, and recommendation engines, where the connections between data points are as meaningful as the data itself.

This chapter will explore the foundational concepts behind graph data models, including how they differ from other data structures, and dive into the core components like nodes, edges, and properties. We will also look at advanced graph querying techniques that enable efficient traversals and complex pattern matching, providing powerful tools for modern data analytics and decision-making.

What Are Graph Data Models: Structure and Key Concepts

Graph data modeling structures data in a way that emphasizes the relationships between entities. Unlike traditional models that store data in tables or rows, graph databases utilize nodes, edges, and properties. Nodes represent entities or objects, such as customers, products, or employees. Edges describe the relationships between these nodes, such as "purchased," "employed by," or "connected to." Properties provide additional details about nodes or edges, such as names, timestamps, or statuses.

The key benefit of graph data models is their ability to represent complex relationships naturally. For example, users can be nodes in a social media platform, and their friendships can be edges, allowing for quick traversal and analysis of direct or indirect relationships. The structure is particularly advantageous for use cases where relationships are highly interconnected, such as fraud detection or recommendation systems, where understanding patterns in data is crucial.

R. L. Steelman, Jr., *Mastering Snowflake DataOps with DataOps.live*, https://doi.org/10.1007/979-8-8688-1754-0_29

Graph databases utilize graph theory principles, where relationships between nodes form graph structures. These structures are typically directed, meaning the relationship flows in one direction, or undirected, where the relationship is bidirectional. For example, an "employee" node might have a directed edge to the "manager" node in a corporate structure, representing the reporting line. At the same time, a "friendship" relationship in a social network might be undirected, indicating mutual relationships.

Another essential concept is the graph traversal process, which allows for efficient querying of data across relationships. Traversal begins at a node and explores its edges to reach other related nodes. It enables complex queries such as finding the shortest path between two nodes or retrieving all entities connected by a specific relationship type. Graph traversal is often more efficient than traditional SQL joins in highly connected datasets, making graph databases ideal for particular applications.

The flexibility of graph data models also comes from their ability to evolve with changing data. As new relationships or entities are introduced, the graph structure can be easily expanded without disrupting existing data. This is particularly useful in dynamic industries like e-commerce or social media, where new connections and behaviors emerge frequently.

Graph data models can be classified into various types, including property graphs, hypergraphs, and RDF graphs. Property graphs, with nodes, edges, and properties, are the most common in commercial applications. In contrast, RDF graphs, which focus on linking data across different domains, are used primarily in semantic web applications. Understanding the differences between these models helps choose the best fit for specific applications.

In addition to structured data, graph models can also handle semi-structured and unstructured data, making them versatile across different data environments. This ability to work with diverse data types makes graph databases a powerful tool in data science and machine learning, particularly in use cases like predictive modeling or anomaly detection, where complex relationships must be analyzed.

Finally, the query language most commonly associated with graph databases is Cypher, used in systems like Neo4j, or Gremlin, used in Apache TinkerPop. These query languages are optimized for graph traversal, enabling developers and data scientists to quickly write queries that reflect the natural relationships between data entities. These languages make it easier to express complex relationships and extract insights that might be difficult to obtain using traditional relational databases.

Designing Graph Databases: Nodes, Edges, and Properties

In graph databases, the three foundational components—nodes, edges, and properties—work together to represent relationships and entities in a structured yet flexible way. Nodes are the entities or objects in the system, such as customers, products, or locations. Each node can hold attributes or properties that provide detailed information about the entity it represents. These properties can include values such as a customer's name, address, or purchase history. In essence, nodes are the data points within the graph.

Edges, also known as relationships, connect the nodes and represent their interactions or associations. Each edge typically has a direction, indicating the flow of a relationship. For example, in a social network, a directed edge may link two user nodes, with the edge indicating a "follows" relationship. Unlike traditional databases, where relationships are typically stored in separate tables, graph databases store relationships directly alongside the entities, making them more efficient for complex queries that involve traversing relationships.

Edges, just like nodes, can also have properties attached to them. These properties provide additional context about the relationship. For instance, in a "purchased" relationship between a customer node and a product node, the edge might contain properties such as the purchase date, quantity, and transaction amount. This allows graph databases to capture the entities involved and the specific details of their interactions, which can be crucial for tasks such as analyzing trends or performing detailed queries.

Properties are the key-value pairs associated with both nodes and edges in graph databases. These properties are used to add metadata or descriptive information that can be queried and manipulated just like the nodes and edges themselves. For example, a product node might contain properties like "price," "category," and "stock availability." In contrast, a relationship edge between a customer and a product could contain properties such as "purchase date," "payment method," and "transaction status." These properties are essential for making the graph more useful for specific business and analytic queries.

Understanding the relationships between the various nodes is crucial when designing a graph database schema. A well-designed graph structure reflects real-world relationships, ensuring the graph provides both efficiency and clarity. For instance, in a recommendation system, the relationships might represent direct interactions, such as purchases or likes, and indirect connections, like "friend of" or "similar to" relationships, which can significantly influence the system's ability to make accurate suggestions.

In practice, nodes and edges should be designed to allow for easy expansion as the data grows. Since graph databases are inherently flexible, it's important to create a design that can scale with changes in business requirements. This often involves considering future use cases, such as new relationship types or additional properties, to ensure the database can accommodate new data needs without significant rework.

A common challenge in graph database design is balancing between generalization and specificity. Nodes and edges should be generalized enough to handle a wide range of queries but specific enough to capture the nuances of the relationships. For instance, in a supply chain management system, you may choose to model both suppliers and customers as "partner" nodes. Still, having distinct nodes for different partner types might be more efficient if unique properties or relationships are associated with each.

Another important aspect of graph database design is understanding the cardinality of relationships, which refers to how many nodes a particular edge can connect. For example, a "manages" relationship between an employee node and a department node may only connect one employee to one department, representing a one-to-one relationship. Conversely, depending on the transaction history, a "purchased" relationship between a customer node and a product node could represent a one-to-many or many-to-many relationship.

Graph database models also require careful consideration of indexing strategies. Since relationships are as meaningful as the data entities themselves, it's vital to ensure that edges are indexed properly to optimize query performance. Many graph database systems allow indexing on properties of nodes and edges, improving search efficiency by providing quick lookups when performing graph traversals or querying specific relationships.

Additionally, it's important to handle data consistency and integrity when designing a graph database. Graphs can become quite complex as new entities and relationships are added. Ensuring that all data adheres to business rules and data integrity constraints, such as preventing orphan nodes or invalid relationships, is a key design consideration. For this reason, most graph databases provide mechanisms for enforcing data integrity, such as referential integrity between nodes and edges or triggers to update relationships when entities change.

Finally, it's beneficial to map out the real-world processes the database aims to represent in designing a graph database schema. Graphs are best suited for complex, interconnected data and are widely used in social networks, recommendation systems, fraud detection, and supply chain management, among others. A strong understanding of the business problem and its data relationships will help ensure the database design remains efficient, scalable, and valuable for analytics. Understanding the data flow and interactions between nodes will provide the necessary insights for crafting a graph schema that effectively captures the business logic.

Advanced Graph Querying: Traversals, Patterns, and Algorithms

Graph querying is one of the most powerful features of graph databases, enabling users to explore relationships between entities and retrieve relevant data efficiently. At its core, graph querying involves traversals, where you follow edges between nodes to analyze the structure and relationships of the graph. These queries often rely on specific patterns or algorithms that allow you to identify meaningful connections between entities, such as finding the shortest path between nodes or detecting clusters of related nodes.

One fundamental concept in advanced graph querying is traversals. A traversal involves starting from one node and following the relationships (edges) to other nodes in the graph. The depth and direction of the traversal depend on the query. For example, in a social network, a simple traversal could start from a person node and explore their immediate connections, then move on to their friends' friends, and so on. This traversal approach allows you to explore different types of relationships and can uncover deeper connections that would be difficult to find in a traditional relational database.

Traversals can be customized with specific rules, such as limiting the number of hops, filtering based on properties, or following particular types of relationships. For example, in a recommendation system, a traversal might start from a user node and explore product nodes with which the user has interacted, then move to other users who interacted with similar products, and recommend items that those users liked. This type of querying leverages the power of graph data models to generate personalized results that would otherwise require complex joins in a relational database.

Pattern matching is another critical technique in advanced graph querying. It involves finding specific structures or motifs within a graph, such as a particular configuration of nodes and edges. Pattern matching can identify specific relationship patterns, like finding cycles, clusters, or communities of interconnected nodes. This is particularly useful in social network analysis, fraud detection, and bioinformatics, where specific relationships between entities must be recognized to draw actionable insights.

Graph algorithms are a natural extension of graph querying, allowing for more sophisticated graph data analysis. One of the most commonly used algorithms in graph databases is the shortest path algorithm, which finds the quickest route between two nodes. This is particularly useful in transportation and logistics systems, where optimal routes between locations need to be determined. The breadth-first search (BFS) and depth-first search (DFS) are two fundamental graph traversal algorithms often employed in these cases.

Another significant graph algorithm is the PageRank algorithm, which Google famously used to rank web pages. PageRank measures the importance of a node in a graph based on the number and quality of its incoming connections. This algorithm can be applied to social networks, citation networks, and recommendation systems to determine the most influential nodes within a network. PageRank recursively calculates scores for each node until the algorithm converges on the optimal ranking.

Community detection algorithms are also widely used in graph querying, particularly in social networks and collaborative filtering systems. These algorithms aim to identify groups of nodes that are more tightly connected to each other than to the rest of the graph. A typical algorithm for community detection is the Louvain method, which efficiently identifies communities in large-scale graphs by optimizing modularity, a measure of how well a graph is divided into communities. This can be invaluable in marketing, customer segmentation, and social media analysis.

Another important graph algorithm is centrality measures, which are used to identify the most critical nodes in a graph. There are several types of centrality, such as degree centrality, closeness centrality, and betweenness centrality. Degree centrality measures the number of direct connections a node has, while closeness centrality evaluates how easily a node can reach other nodes in the graph. Betweenness centrality measures how often a node acts as a bridge in the shortest path between two other nodes. These measures are valuable in network analysis, epidemiology, and social influence studies.

Graph clustering is another advanced graph query technique that groups nodes into clusters based on their connectivity. Clustering allows you to identify groups of densely connected nodes and separate them from other parts of the graph. In marketing, for instance, this can be used to identify groups of customers with similar buying patterns or interests. The k-means clustering algorithm and hierarchical clustering are commonly applied to graph data to partition nodes into meaningful groups for further analysis.

One of the challenges in graph querying is ensuring that queries perform efficiently, especially as the graph grows larger. Graph databases often use indexing strategies to optimize query performance, like indexing nodes and edges based on specific properties. Additionally, caching frequently accessed parts of the graph can significantly improve query response times. As graph databases scale, maintaining query performance becomes an ongoing task that requires careful design and tuning.

For large-scale graph queries, distributed graph databases like Apache TinkerPop and Neo4j offer solutions that allow you to process graph data across multiple machines. These distributed systems ensure that graph traversals, pattern matching, and graph algorithms can be executed in parallel, providing scalability and fault tolerance. In cloud environments, graph data models are often deployed using serverless architectures, where resources are allocated dynamically to handle fluctuating query loads.

Lastly, hybrid models are becoming increasingly popular in advanced graph querying, where graph data is combined with other data models, such as relational or document-based models. In such systems, graph queries may be used to explore relationships between entities stored in relational databases or data lakes. This approach leverages the strengths of graph data models for analyzing relationships while also enabling the power of traditional databases for handling structured data. Integrating graph data with other data models opens up new data analysis and processing possibilities, making it more adaptable to modern data environments.

Conclusion

Graph data modeling represents a powerful paradigm that enables businesses to capture and analyze complex relationships in their data. By modeling real-world interactions, graph databases offer a unique advantage in understanding intricate connections that traditional relational databases might miss. By leveraging nodes, edges, and properties, graph models allow for more intuitive data analysis, particularly in domains like social networks, recommendation systems, and fraud detection.

Advanced querying techniques, such as traversals and graph algorithms, further enhance the power of graph data models. These techniques allow users to explore vast data networks efficiently, uncovering key relationships and patterns in decision-making processes. Using algorithms like PageRank or community detection will enable organizations to identify influential entities or tightly knit groups within their data, providing valuable insights that drive strategic business actions.

Integrating graph data models with other systems, such as relational databases or data lakes, broadens the scope of data analytics even further. By combining the strengths of various data models, organizations can maintain the robustness of traditional systems while leveraging the flexibility and power of graph queries. Hybrid models provide a comprehensive approach that ensures a wide range of use cases can be effectively addressed, whether it's exploring relationships, performing large-scale analyses, or maintaining structured data.

Graph databases and querying techniques have proven to be essential for handling large-scale, interconnected datasets, which are becoming increasingly common in today's data-driven world. As data continues to grow in complexity and volume, the ability to traverse and analyze relationships within it becomes a critical capability. Organizations harnessing graph data models' power can unlock deeper insights, making more informed and timely decisions.

As the field of graph data modeling continues to evolve, new algorithms, technologies, and best practices will emerge, further refining how we capture, analyze, and interpret relationships. The increasing adoption of graph databases, coupled with advancements in processing power and cloud computing, will drive innovation in data analytics. Embracing this paradigm opens up opportunities for businesses to stay competitive by making data-driven decisions based on the rich, interconnected data that lies at the heart of modern industries.

Index

A

Agile data development, 5, 6
Agile data modeling, 275
AI, *see* Artificial intelligence (AI)
AI-driven security tools, 18
AI-powered analytics tools, 18
Amazon SageMaker, 88
Apache Airflow, 55, 239
Apache Kafka, 320
Apache Spark, 250
Apache TinkerPop, 328, 333
Artificial intelligence (AI), 14, 17
Automated data testing, 40
Automated Machine Learning (AutoML), 47
Automating data pipelines, 68, 69
AutoML, *see* Automated Machine Learning (AutoML)
AWS EventBridge, 320
AWS Kinesis, 323
Azure Machine Learning, 88

B

BCNF, *see* Boyce-Codd Normal Form (BCNF)
BFS, *see* Breadth-first search (BFS)
BI, *see* Business intelligence (BI)
Bitbucket, 96, 101
Boyce-Codd Normal Form (BCNF), 313
Branching strategies, 193
Breadth-first search (BFS), 332
Build Only Changed Models, 241, 242
Business intelligence (BI), 173, 255

C

Cardinality, 313
Centralized version control system (CVCS), 94
CircleCI, 147, 150
Cloud-based data lakes, 238
Cloud-based DataOps platforms, 14
Cloud computing, 6
Cloud-native technologies, 11
Collaboration and self-service, 41
Common Table Expressions (CTEs), 138, 141
Continuous integration and continuous delivery (CI/CD), 7, 9, 13, 33, 164, 172
Continuous integration and continuous deployment (CI/CD), 44, 51, 108, 169, 201
Continuous integration and continuous deployment (CI/CD) pipelines, 1, 47
Cross-functional teams
  advantage, 66
  agile data operations, 66
  collaboration tools, 67
  communication and collaboration, 66
  fragmentation, 65

R. L. Steelman, Jr., *Mastering Snowflake DataOps with DataOps.live*, https://doi.org/10.1007/979-8-8688-1754-0

CTEs, *see* Common Table
Expressions (CTEs)
CVCS, *see* Centralized version control
system (CVCS)

D

Database management
best practices and
troubleshooting, 229–231
cross-database workflows, 231
DataOps.live, 226
optimizing cross-database workflows,
228, 229
setting up multiple databases, 227
Databricks, 250
Data Build Tool (DBT), 215, 219
artifacts, 157–159
data pipelines, 125, 126
datasets and teams, 161
data transformations, 140–143
definition, 121
developing and running, 129–131
documentation, 124, 131–134
enterprise-level deployments, 160, 161
ETL tools, 123, 124
features, 162
incremental loading, 153–155
large-scale DBT projects, 149–151
macros and Jinja, 143–146
models for performance, 138–140
run-time, 151–153
setting up, 127–129
SQL-based models, 122
testing, 122, 131–134, 146–149
version control and collaboration,
134–136, 155, 156
Data engineers, 66
Data governance, 13, 15, 251
Data governance and compliance, 71, 72
Data lineage, 16, 72–74
Data mart, 266
DataOps, 1
AI/ML integration, 17–19
automation and CI/CD pipelines, 9–12
big data/agile data development, 4–6
cloud and hybrid architecture, 74, 75
data pipelines, 4
data quality, 15, 17
DevOps, 2, 3, 7–9
emergence, 12–14
pipeline, 199–202
principles
automation and
orchestration, 23, 24
benefits, 27–30
CI/CD, 20, 21
collaboration and
communication, 21, 22
data quality and governance,
24, 25
scalability and flexibility, 26, 27
DataOps.live, 225, 226
accounts, 177, 178
best practice, 186
data pipelines and jobs, 185
data workflow, 185
environment, 184
features, 182
groups and subgroups, 178, 179
orchestrators, 233
pipelines and jobs, 181, 182
projects, 179, 180
REST API, 243–245
scaling and monitoring, 186
Snowflake Marketplace, 183, 184

DataOps.live platform
- agile methodologies, 163
- data environments, 163
- development principles, 165, 166
- features/capabilities, 164, 165
- practices, 171, 172
- principles, 169–171, 175
- Snowflake Data Cloud, 167, 168

Data quality, 69–71
DataRobot, 88
Data security, 15
Data validation, 253
Data Vault, 311
DataVault 2.0, 173, 174, 317
- best practices, 290–292
- components, 276
  - hubs, 280–282
  - links, 280, 283, 284
  - satellites, 285, 286
- healthcare industry, 292–295
- implementation, 289, 290
- large-scale data environments, 275
- modern data environments, 287–289
- principles, 277–279
- real-time data integration, 276

Data Version Control (DVC), 67
Data warehouses, 323
Data.World Orchestrator, 234–236
DBT, *see* Data Build Tool (DBT)
Debian-based systems, 105
Depth-first search (DFS), 332
DevOps, 2, 7
DFS, *see* Depth-first search (DFS)
Distributed version control system (DVCS), 94
DVC, *see* Data Version Control (DVC)

E, F

EDA, *see* Event-driven architecture (EDA)
Effective environment management
- branching strategies, 193–196
- creating DataOps environments, 191–193
- data pipelines, 189, 197
- segregation and configuration, 190, 191
- types of environments, 189, 190

EHR, *see* Electronic health records (EHR)
Electronic health records (EHR), 88
End-to-end governance framework, 172
Entity-Relationship (ER) modeling, 311
- attributes, 312
- cardinality, 313, 314
- data ecosystems, 311
- modern data architecture, 316, 317
- normalization, 312
- relationships, 312
- temporal data, 315

Environment management, 38
ETL, *see* Extract, Transform, Load (ETL)
Event-driven architecture (EDA), 319
Event-driven data models, 320
- advantage, 319
- data consistency, 326
- data warehouses/data lakes, 323, 324
- designing, 321, 322
- implementation, 324, 325
- principles, 320, 321

Extract, Transform, Load (ETL), 33, 34, 75, 125, 179, 199, 234, 265, 269, 270
- advantage, 34
- CI/CD, 35, 36
- data management, 34

G

Git, 67, 202
- advanced capabilities, 113
- aliases, 109
- best practices, 116–118
- branching, 113–115
- collaboration, 108
- collaborative development, 98, 99
- definition, 95, 96
- deployment processes, 99
- flags, 102
- installation, 103–106
- merging, 115, 116
- principles and workflow, 96, 97
- remotes, 111
- repositories, 101, 102
- tagging, 110
- troubleshooting, 118, 119
- version control, 93–95, 101, 112
- working, 107, 108

git fetch, 114
"GitFlow" model, 97
GitHub, 96, 101, 150
GitLab, 55, 96, 101
git push, 118
git rebase, 114, 118
Graph algorithms, 333
Graph clustering, 332
Graph data modeling
- databases, 334
- designing, 329, 330
- hybrid models, 334
- querying, 331–333
- structures, 327, 328
- structuring and querying data, 327

Graphical user interfaces (GUIs), 106
Graph querying, 331
Gremlin, 328
GUIs, *see* Graphical user interfaces (GUIs)

H

Hubs, 276, 280, 295

I

is_incremental() function, 151, 153

J

Jenkins, 55, 147, 150, 243
Jinja templating, 142

K

Kafka, 323
Key performance indicators (KPIs), 48, 55, 269, 284, 308
Kimball architecture
- building data marts, 269, 270
- data mart, 266
- designing data warehouses, 265
- designing dimensional models, 267, 268
- ETL, 270, 271
- practices and optimization strategies, 272, 273

Kimball methodology, 175
Kimball modeling, 317
KPIs, *see* Key performance indicators (KPIs)
Kubeflow, 55

L

Learning management systems (LMS), 90
LMS, *see* Learning management systems (LMS)

M

Machine learning (ML), 3, 14, 17, 31
Machine Learning Operations (MLOps), 3, 42
  automation, 44, 54–56
  case studies, 60–62
  challenges, 50–52
  data pipelines, 43
  governance/compliance, 59, 60
  lifecycle, 45–48
  model deployment, 56, 57
  monitoring and maintenance, 45, 58
  principles, 49
  real-world applications, 60–62
  reproducibility, 44
  versioning, 52–54
MATE, *see* Modelling and Transformation Engine (MATE)
Medallion, 173, 175, 311
  combining with DataVault 2.0, and Kimball
    best practices, 302, 303
    end-to-end data pipeline, 297
    finance industry, 307–310
    Gold layer, 298, 300, 301
    managing and scaling, 306
    optimizing data flow, 303–305
    Silver layer, 298–300
Medallion architecture
  advantage, 250
  best practices and optimization strategies, 257–259
  Bronze layer, 250, 251
  data pipelines, 248
  Gold layer, 254–256
  implementation, 256, 257
  large data environments, 260
  layered approach, 247
  layers, 248, 249
  modern data ecosystems, 247
  Silver layer, 252, 253
  solar energy, 261–264
Metadata management, 16
ML, *see* Machine learning (ML)
MLOps, *see* Machine Learning Operations (MLOps)
Model deployment, 56
Modelling and Transformation Engine (MATE), 166
  best practices, 222–224
  building models, 217–219
  collaboration, 224
  data management processes, 215
  DBT packages, 215, 216, 219–221
  elements, 217
  macros, 221, 222
  YAML and SQL templates, 217
MuleSoft, 173
Multi-Cluster Shared Data Architecture, 82

N, O

Natural language processing (NLP), 294
Neo4j, 328, 333
NLP, *see* Natural language processing (NLP)
Normalization, 313
NoSQL systems, 238

P, Q

PageRank algorithm, 332

R

RBACs, *see* Role-based access controls (RBACs)
Red Hat-based systems, 105
ref() function, 124, 125, 153
Role-based access controls (RBACs), 79, 80, 86, 210, 235

S

SCDs, *see* Slowly changing dimensions (SCDs)
SIS, *see* Student information systems (SIS)
Slowly changing dimensions (SCDs), 315
Snowflake, 250
  architecture, 77
  benefits, data warehousing, 80–82
  data ecosystem, 77
  data security and compliance, 86
  data sharing, 80, 84, 85
  features, 78, 79
  industry-specific applications, 91
  multi-cloud capabilities, 91
  multi-cluster shared data architecture, 82, 83
  performance optimization, 85, 86
  semi-structured data, 79
  storage and compute separation, 83, 84
  third-party tools and platforms, 87
  use cases and industry applications, 88–90
  virtual warehouses, 79
Snowflake Data Cloud, 167
Snowflake Object Lifecycle Engine (SOLE), 166
  advanced configuration options, 211, 212
  best practices, 209, 210
  data products, 204–207
  machine learning models, 213
  modular/self-describing object configuration, 207–209
  organizing data objects, 203
  YAML-based configuration files, 203
SOLE, *see* Snowflake Object Lifecycle Engine (SOLE)
SQL-based applications, 87
SQL-based approach, 216
Student information systems (SIS), 90

T, U

Tags, 102
Testing, 70
Traditional database engineering, 1
Traversals, 333
Tried-and-true method, 273
#TrueDataOps, 1, 36–39

V, W, X, Y

Vaults, 181
VaultSpeed, 236–239
VCS, *see* Version control systems (VCS)
Version control, 72–74, 93–95
Version-controlled SQL-based transformations, 121
Version control systems (VCS), 242

Z

Zero Copy Cloning, 38

GPSR Compliance
The European Union's (EU) General Product Safety Regulation (GPSR) is a set of rules that requires consumer products to be safe and our obligations to ensure this.

If you have any concerns about our products, you can contact us on

ProductSafety@springernature.com

In case Publisher is established outside the EU, the EU authorized representative is:

Springer Nature Customer Service Center GmbH
Europaplatz 3
69115 Heidelberg, Germany

www.ingramcontent.com/pod-product-compliance
Lightning Source LLC
LaVergne TN
LVHW081347060526
838201LV00050B/1732